WEST-E Physics
0265 Teacher Certification Exam

By: Sharon Wynne, M.S
Southern Connecticut State University

"And, while there's no reason yet to panic, I think it's only prudent that we make preparations to panic."

XAMonline, INC.
Boston

Copyright © 2008 XAMonline, Inc.
All rights reserved. No part of the material protected by this copyright notice may be reproduced or utilized in any form or by any means, electronic or mechanical, including photocopying, recording or by any information storage and retrievable system, without written permission from the copyright holder.

To obtain permission(s) to use the material from this work for any purpose including workshops or seminars, please submit a written request to:

> XAMonline, Inc.
> 21 Orient Ave.
> Melrose, MA 02176
> Toll Free 1-800-509-4128
> Email: info@xamonline.com
> Web www.xamonline.com
> Fax: 1-718-662-9268

Library of Congress Cataloging-in-Publication Data

Wynne, Sharon A.
    Physics 0265: Teacher Certification / Sharon A. Wynne. -2nd ed.
    ISBN 978-1-58197-043-2
    1. Physics 0265.   2. Study Guides.   3. WEST
    4. Teachers' Certification & Licensure.   5. Careers

Disclaimer:

The opinions expressed in this publication are the sole works of XAMonline and were created independently from the National Education Association, Educational Testing Service, or any State Department of Education, National Evaluation Systems or other testing affiliates.

Between the time of publication and printing, state specific standards as well as testing formats and website information may change that is not included in part or in whole within this product. Sample test questions are developed by XAMonline and reflect similar content as on real tests; however, they are not former tests. XAMonline assembles content that aligns with state standards but makes no claims nor guarantees teacher candidates a passing score. Numerical scores are determined by testing companies such as NES or ETS and then are compared with individual state standards. A passing score varies from state to state.

Printed in the United States of America   œ-1

WEST-E: Physics 0265
ISBN: 978-1-58197-043-2

TEACHER CERTIFICATION STUDY GUIDE

Table of Contents

Pg.

Domain I. Mechanics

Skill 1. Vectors (properties; addition and subtraction) 1

Skill 2. Vector multiplication (dot and cross product) 3

Skill 3. Motion along a straight line (displacement, velocity, acceleration) .. 4

Skill 4. Motion in two dimensions (projectile motion, uniform circular motion) 6

Skill 5. Reference frames and relative motion (relative velocity, Galilean relativity) 8

Skill 6. Force and Newton's laws of motion (Newton's first law, inertia, inertial reference frames, Newton's second law, force and acceleration, addition of forces, balanced versus unbalanced forces, Newton's third law, action-reaction forces, weight and mass) 9

Skill 7. Friction (static friction, kinetic friction, rolling friction) 11

Skill 8. Equilibrium of forces 12

Skill 9. Equilibrium of moments (torques) 13

Skill 10. Dynamics of uniform circular motion 15

Skill 11. Work, energy, and power (relationship between work and kinetic energy, work done by a variable force) 16

Skill 12. Conservation of energy (potential energy, conservative and nonconservative forces) 19

Skill 13. Simple harmonic motion and oscillations (Hooke's law, graphical and mathematical representations, energy considerations, pendulums, springs) 20

Skill 14. Linear momentum and impulse (momentum-impulse relationship, conservation of linear momentum; elastic and inelastic collisions) 21

Skill 15. Rigid body motion (angular velocity and angular acceleration, angular momentum, moment of inertia, torque, and center of mass, conservation of angular momentum, rotational kinetic energy) 24

Skill 16. Mass-energy relationships (conservation of mass-energy) 27

PHYSICS

TEACHER CERTIFICATION STUDY GUIDE

Skill 17.	Newton's law of universal gravitation and orbital motion (motion of satellites)	27
Skill 18.	Kepler's laws (law of orbits (first law), law of areas (second law), law of periods (third law))	28
Skill 19.	Fluids (density and pressure, ideal fluids at rest, Pascal's law, Archimedes' principle and buoyant forces)	31
Skill 20.	Ideal fluids in motion (Bernoulli's principle, streamlines, equation of continuity)	33

Domain II. Electricity and Magnetism

Skill 1.	Electric forces and Coulomb's law	36
Skill 2.	Electric fields, Gauss's law, electric potential energy, electric potential, and potential difference	38
Skill 3.	Conductors, insulators and semiconductors (charging by friction, conduction and induction)	40
Skill 4.	Capacitance and dielectrics	42
Skill 5.	Conductors, insulators, and semiconductors as used in circuits	44
Skill 6.	Sources of EMF (batteries, photocells, generators)	44
Skill 7.	Current and resistance (Ohm's law, resistivity)	45
Skill 8.	Capacitance and inductance	47
Skill 9.	Energy and power	47
Skill 10.	Analyzing circuits (series and parallel circuits using Ohm's law or Kirchhoff's rules, resistors and capacitors in series or parallel, internal resistance, RC circuits)	48
Skill 11.	Power in alternating-current circuits (average power and energy transmission)	52
Skill 12.	Measurement of potential difference, current, resistance, and capacitance (ammeter, galvanometer, voltmeter and potentiometer)	53
Skill 13.	Magnets, magnetic fields, and magnetic forces (magnetic dipoles and materials, forces on a charged particle moving in a magnetic and/or electric field (Lorentz force law, cyclotron, mass spectrometer) forces or torques on current carrying conductors in magnetic fields)	54
Skill 14.	Magnetic flux (Gauss's law of magnetism)	57
Skill 15.	Magnetic fields produced by currents (Biot-Savart law, Ampere's law, magnetic field of a wire, magnetic field of a solenoid, displacement current)	58

TEACHER CERTIFICATION STUDY GUIDE

Skill 16.	Electromagnetic induction (magnetic flux, Lenz's law, Faraday's law, transformers, generators and motors)	60

Domain III. Optics and Waves

Skill 1.	Wave speed, amplitude, wavelength, and frequency	63
Skill 2.	Inverse square law for intensity	65
Skill 3.	Reflection, refraction, absorption, transmission, and scattering (Snell's law, Rayleigh scattering)	65
Skill 4.	Transverse and longitudinal waves and their properties (Doppler effect, resonance and natural frequencies, polarization)	66
Skill 5.	Sound waves (pitch and loudness, air columns and standing waves open at both ends and closed at one end, harmonics, beats)	67
Skill 6.	Electromagnetic spectrum (frequency regions, color)	69
Skill 7.	Principle of linear superposition and interference (diffraction, dispersion, beats and standing waves, interference in thin films and Young's double-slit experiment)	70
Skill 8.	Reflection and refraction (Snell's law, total internal reflection, fiber optics)	73
Skill 9.	Thin lenses	75
Skill 10.	Plane and spherical mirrors	77
Skill 11.	Prisms	79
Skill 12.	Optical instruments (simple magnifier, microscope, telescope)	80

Domain IV. Heat and Thermodynamics

Skill 1.	Measurement of heat and temperature (temperature scales)	83
Skill 2.	Thermal expansion	84
Skill 3.	Thermocouples	85
Skill 4.	Heat capacity and specific heat, Latent heat of phase change (heat of fusion, heat of vaporization)	86
Skill 5.	Transfer of heat (conduction, convection and radiation)	89
Skill 6.	Kinetic molecular theory (ideal gas laws)	91
Skill 7.	First law of thermodynamics (internal energy, energy conservation)	94
Skill 8.	Thermal processes involving pressure, volume and temperature	95
Skill 9.	Second law of thermodynamics (entropy and disorder, reversible and irreversible processes, spontaneity, heat engines, Carnot cycle, efficiency)	97

PHYSICS

TEACHER CERTIFICATION STUDY GUIDE

Skill 10.	Third law and zeroth law of thermodynamics (absolute zero of temperature, law of equilibrium)	99
Skill 11.	Energy and energy transformations (kinetic, potential, mechanical, sound, magnetic, electrical, light, heat, nuclear, chemical)	100

Domain V.   Modern Physics, Atomic and Nuclear Structure

Skill 1.	Nature of the atom (Rutherford scattering, atomic models, Bohr model, atomic spectra)	101
Skill 2.	Atomic and nuclear structure (electrons, protons and neutrons; electron arrangement, isotopes, hydrogen atom energy levels, nuclear forces and binding energy)	103
Skill 3.	Radioactivity (radioactive decay, half life, isotopes, decay processes, alpha decay, beta decay, gamma decay, artificial radioactivity)	106
Skill 4.	Elementary particles (ionizing radiation)	109
Skill 5.	Organization of matter (elements, compounds, solutions, and mixtures)	111
Skill 6.	Physical properties of matter (phase changes, states of matter)	112
Skill 7.	Nuclear energy (fission and fusion, nuclear reactions and their products)	114
Skill 8.	Blackbody radiation and photoelectric effect	117
Skill 9.	de Broglie's hypothesis and wave-particle duality	120
Skill 10.	Special relativity (Michelson-Morley experiment (ether and the speed of light), simultaneity, Lorentz transformations, time dilation, length contraction, velocity addition)	121

Domain VI.   History and Nature of Science; Science, Technology, and Social Perspectives (STS)

Skill 1.	Scientific method of inquiry (formulating problems, formulating and testing hypotheses, making observations, developing generalizations)	123
Skill 2.	Science process skills (observing, hypothesizing, ordering, categorizing, comparing, inferring, applying, communicating)	126
Skill 3.	Distinguish among hypotheses, assumptions, models, laws, and theories	127
Skill 4.	Experimental design (Data collection, interpretation and presentation, significance of controls)	130
Skill 5.	Integrate the overarching concepts of science	131

TEACHER CERTIFICATION STUDY GUIDE

Skill 6.	Historical roots of the physical sciences and the contributions made by major historical figures to the physical sciences	132
Skill 7.	Scientific knowledge is subject to change	135
Skill 8.	Scientific measurement and notation systems	136
Skill 9.	Processes involved in scientific data collection and manipulation (organization of data, significant figures, linear regression)	139
Skill 10.	Interpret and draw conclusions from data, including those presented in tables, graphs, and charts	142
Skill 11.	Analyze errors in data that is presented (sources of error, accuracy, precision)	145
Skill 12.	Safety procedures involved in the preparation, storage, use, and disposal of laboratory and field materials	147
Skill 13.	Identify appropriate use, calibration procedures, and maintenance procedures for laboratory and field equipment	148
Skill 14.	Preparation of reagents, materials, and apparatus for classroom use	150
Skill 15.	Knowledge of safety and emergency procedures for the science classroom and laboratory	150
Skill 16.	Knowledge of the legal responsibilities of the teacher in the science classroom	152
Skill 17.	Impact of science and technology on the environment and human affairs	152
Skill 18.	Issues associated with energy production, transmission, management, and use (including nuclear waste removal and transportation)	153
Skill 19.	Issues associated with the production, storage, use, management, and disposal of consumer products	154
Skill 20.	Issues associated with the management of natural resources	155
Skill 21.	Applications of science and technology in daily life	156
Skill 22.	Social, political, ethical, and economic issues arising from science and technology	157

Sample Test 159

Answer Key 188

Rigor Table 189

Rationales with Sample Questions 190

TEACHER CERTIFICATION STUDY GUIDE

Great Study and Testing Tips!

What to study in order to prepare for the subject assessments is the focus of this study guide but equally important is *how* you study.

You can increase your chances of truly mastering the information by taking some simple, but effective steps.

Study Tips:

1. Some foods aid the learning process. Foods such as milk, nuts, seeds, rice, and oats help your study efforts by releasing natural memory enhancers called CCKs (*cholecystokinin*) composed of *tryptophan*, *choline*, and *phenylalanine*. All of these chemicals enhance the neurotransmitters associated with memory. Before studying, try a light, protein-rich meal of eggs, turkey, and fish. All of these foods release the memory enhancing chemicals. The better the connections, the more you comprehend.

Likewise, before you take a test, stick to a light snack of energy boosting and relaxing foods. A glass of milk, a piece of fruit, or some peanuts all release various memory-boosting chemicals and help you to relax and focus on the subject at hand.

2. Learn to take great notes. A by-product of our modern culture is that we have grown accustomed to getting our information in short doses (i.e. TV news sound bites or USA Today style newspaper articles.)

Consequently, we've subconsciously trained ourselves to assimilate information better in neat little packages. If your notes are scrawled all over the paper, it fragments the flow of the information. Strive for clarity. Newspapers use a standard format to achieve clarity. Your notes can be much clearer through use of proper formatting. A very effective format is called the *"Cornell Method."*

Take a sheet of loose-leaf lined notebook paper and draw a line all the way down the paper about 1-2" from the left-hand edge.

Draw another line across the width of the paper about 1-2" up from the bottom. Repeat this process on the reverse side of the page.

Look at the highly effective result. You have ample room for notes, a left hand margin for special emphasis items or inserting supplementary data from the textbook, a large area at the bottom for a brief summary, and a little rectangular space for just about anything you want.

TEACHER CERTIFICATION STUDY GUIDE

3. Get the concept then the details. Too often we focus on the details and don't gather an understanding of the concept. However, if you simply memorize only dates, places, or names, you may well miss the whole point of the subject.

A key way to understand things is to put them in your own words. If you are working from a textbook, automatically summarize each paragraph in your mind. If you are outlining text, don't simply copy the author's words.

Rephrase them in your own words. You remember your own thoughts and words much better than someone else's, and subconsciously tend to associate the important details to the core concepts.

4. Ask Why? Pull apart written material paragraph by paragraph and don't forget the captions under the illustrations.

Example: If the heading is "Stream Erosion", flip it around to read "Why do streams erode?" Then answer the questions.

If you train your mind to think in a series of questions and answers, not only will you learn more, but it also helps to lessen the test anxiety because you are used to answering questions.

5. Read for reinforcement and future needs. Even if you only have 10 minutes, put your notes or a book in your hand. Your mind is similar to a computer; you have to input data in order to have it processed. *By reading, you are creating the neural connections for future retrieval.* The more times you read something, the more you reinforce the learning of ideas.

Even if you don't fully understand something on the first pass, *your mind stores much of the material for later recall.*

6. Relax to learn so go into exile. Our bodies respond to an inner clock called biorhythms. Burning the midnight oil works well for some people, but not everyone.

If possible, set aside a particular place to study that is free of distractions. Shut off the television, cell phone, and pager and exile your friends and family during your study period.

If you really are bothered by silence, try background music. Light classical music at a low volume has been shown to aid in concentration over other types. Music that evokes pleasant emotions without lyrics is highly suggested. Try just about anything by Mozart. It relaxes you.

PHYSICS

7. Use arrows not highlighters. At best, it's difficult to read a page full of yellow, pink, blue, and green streaks. Try staring at a neon sign for a while and you'll soon see that the horde of colors obscure the message.

A quick note, a brief dash of color, an underline, and an arrow pointing to a particular passage is much clearer than a horde of highlighted words.

8. Budget your study time. Although you shouldn't ignore any of the material, *allocate your available study time in the same ratio that topics may appear on the test.*

TEACHER CERTIFICATION STUDY GUIDE

Testing Tips:

1. Get smart, play dumb. **Don't read anything into the question.** Don't make an assumption that the test writer is looking for something else than what is asked. Stick to the question as written and don't read extra things into it.

2. Read the question and all the choices *twice* before answering the question. You may miss something by not carefully reading, and then re-reading both the question and the answers.

If you really don't have a clue as to the right answer, leave it blank on the first time through. Go on to the other questions, as they may provide a clue as to how to answer the skipped questions.

If later on, you still can't answer the skipped ones . . . ***Guess.*** The only penalty for guessing is that you *might* get it wrong. Only one thing is certain; if you don't put anything down, you will get it wrong!

3. Turn the question into a statement. Look at the way the questions are worded. The syntax of the question usually provides a clue. Does it seem more familiar as a statement rather than as a question? Does it sound strange?

By turning a question into a statement, you may be able to spot if an answer sounds right, and it may also trigger memories of material you have read.

4. Look for hidden clues. It's actually very difficult to compose multiple-foil (choice) questions without giving away part of the answer in the options presented.

In most multiple-choice questions you can often readily eliminate one or two of the potential answers. This leaves you with only two real possibilities and automatically your odds go to Fifty-Fifty for very little work.

5. Trust your instincts. For every fact that you have read, you subconsciously retain something of that knowledge. On questions that you aren't really certain about, go with your basic instincts. **Your first impression on how to answer a question is usually correct.**

6. Mark your answers directly on the test booklet. Don't bother trying to fill in the optical scan sheet on the first pass through the test.

Just be very careful not to miss-mark your answers when you eventually transcribe them to the scan sheet.

7. Watch the clock! You have a set amount of time to answer the questions. Don't get bogged down trying to answer a single question at the expense of 10 questions you can more readily answer.

PHYSICS

DOMAIN I. MECHANICS

Skill 1   Vectors (properties; addition and subtraction)

Vector space is a collection of objects that have **magnitude** and **direction**. They may have mathematical operations, such as addition, subtraction, and scaling, applied to them. Vectors are usually displayed in boldface or with an arrow above the letter. They are usually shown in graphs or other diagrams as arrows. The length of the arrow represents the magnitude of the vector while the direction in which the arrow points shows the vector direction.

To **add two vectors** graphically, the base of the second vector is drawn from the point of the first vector as shown below with vectors **A** and **B**. The sum of the vectors is drawn as a dashed line, from the base of the first vector to the tip of the second. As illustrated, the order in which the vectors are connected is not significant as the endpoint is the same graphically whether **A** connects to **B** or **B** connects to **A**. This principle is sometimes called the parallelogram rule.

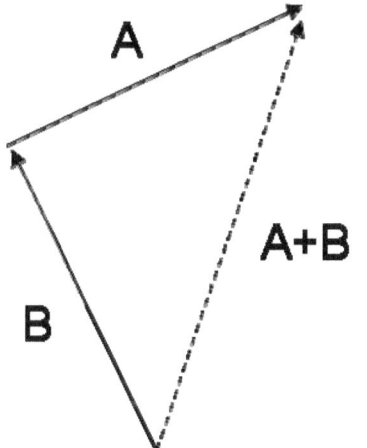 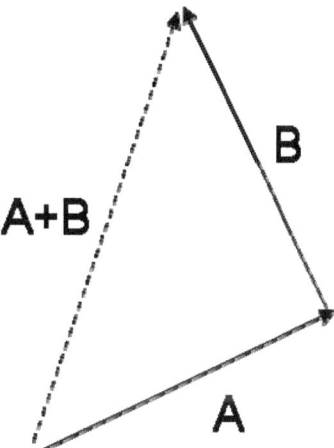

If more than two vectors are to be combined, additional vectors are simply drawn in accordingly with the sum vector connecting the base of the first to the tip of the final vector.

Subtraction of two vectors can be geometrically defined as follows. To subtract **A** from **B**, place the ends of **A** and **B** at the same point and then draw an arrow from the tip of **A** to the tip of **B**. That arrow represents the vector **B-A**, as illustrated below:

PHYSICS

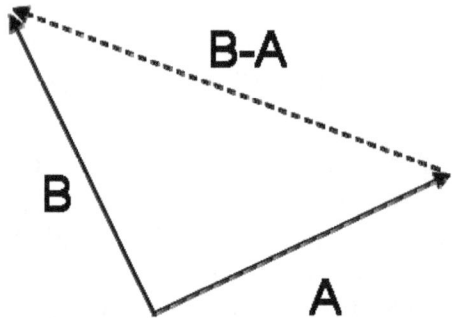

To add two vectors without drawing them, the vectors must be broken down into their orthogonal components using sine, cosine, and tangent functions. Add both x components to get the total x component of the sum vector, then add both y components to get the y component of the sum vector. Use the Pythagorean Theorem and the three trigonometric functions to the get the size and direction of the final vector.

Example: Here is a diagram showing the x and y-components of a vector D1:

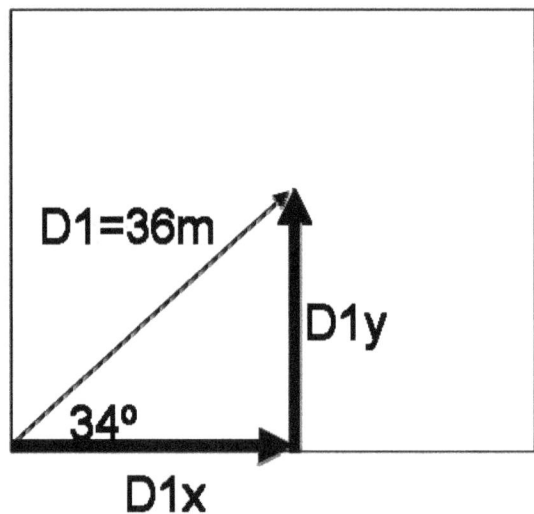

Notice that the x-component D1x is adjacent to the angle of 34 degrees.

Thus D1x=36m (cos34) =29.8m

The y-component is opposite to the angle of 34 degrees.

Thus D1y =36m (sin34) = 20.1m

A second vector D2 is broken up into its components in the diagram below using the same techniques. We find that D2y=9.0m and D2x=-18.5m.

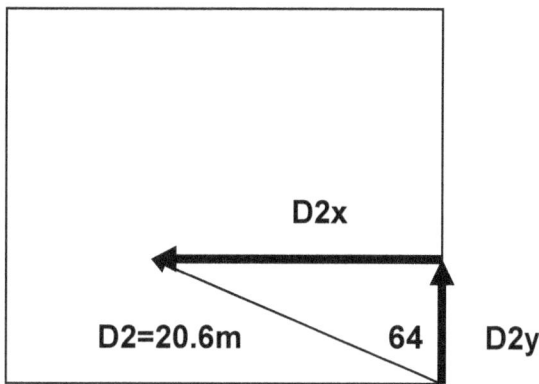

Next we add the x components and the y components to get

DTotal x = 11.3 m   and   DTotal y = 29.1 m

Now we have to use the Pythagorean theorem to get the total magnitude of the final vector. And the arctangent function to find the direction. As shown in the diagram below.

DTotal = 31.2m

$\tan \theta$ = DTotal y / DTotal x = 29.1m / 11.3 = 2.6 θ = 69 degrees

Skill 2   Vector multiplication (dot and cross product)

The dot product is also known as the scalar product. This is because the dot product of two vectors is not a vector, but a scalar (i.e., a real number without an associated direction). The definition of the dot product of the two vectors **a** and **b** is:

$$a \bullet b = \sum_{i=1}^{n} a_i b_i = a_1 b_1 + a_2 b_2 + ... + a_n b_n$$

The following is an example calculation of the dot product of two vectors:

$[1\ 3\ -5] \cdot [4\ -2\ -2] = (1)(4) + (3)(-2) + (-5)(-2) = 8$

Note that the product is a simple scalar quantity, not a vector. The dot product is commutative and distributive.

Unlike the dot product, the cross product does return another vector. The vector returned by the cross product is orthogonal to the two original vectors. The cross product is defined as:

$$\mathbf{a} \times \mathbf{b} = n\,|\mathbf{a}|\,|\mathbf{b}|\sin\theta$$

where n is a unit vector perpendicular to both **a** and **b** and θ is the angle between **a** and **b**. In practice, the cross product can be calculated as explained below:

Given the orthogonal unit vectors **i**, **j**, and **k**, the vector **a** and **b** can be expressed:

$$\mathbf{a} = a_1\mathbf{i} + a_2\mathbf{j} + a_3\mathbf{k}$$
$$\mathbf{b} = b_1\mathbf{i} + b_2\mathbf{j} + b_3\mathbf{k}$$

Then we can calculate that

$$\mathbf{a} \times \mathbf{b} = \mathbf{i}(a_2 b_3) + \mathbf{j}(a_3 b_1) + \mathbf{k}(a_1 b_2) - \mathbf{i}(a_3 b_2) - \mathbf{j}(a_1 b_3) - \mathbf{k}(a_2 b_1)$$

The cross product is anticommutative (that is, $\mathbf{a} \times \mathbf{b} = -\mathbf{b} \times \mathbf{a}$) and distributive over addition.

Skill 3   Motion along a straight line (displacement, velocity, acceleration)

Kinematics is the part of mechanics that seeks to understand the motion of objects, particularly the relationship between position, velocity, acceleration and time.

$X<0$ $\qquad\qquad\qquad$ $X=0$ $\qquad\qquad\qquad\qquad$ $X>0$

The above figure represents an object and its displacement along one linear dimension.

First we will define the relevant terms:

1. Position or Distance is usually represented by the variable x. It is measured relative to some fixed point or datum called the origin in linear units, meters, for example.

2. Displacement is defined as the change in position or distance which an object has moved and is represented by the variables D, d or Δx. Displacement is a vector with a magnitude and a direction.

PHYSICS

3. Velocity is a vector quantity usually denoted with a V or v and defined as the rate of change of position. Typically units are distance/time, m/s for example. Since velocity is a vector, if an object changes the direction in which it is moving it changes its velocity even if the speed (the scalar quantity that is the magnitude of the velocity vector) remains unchanged.

i) Average velocity: $\vec{v} \equiv \dfrac{\Delta d}{\Delta t} = d_1 - d_0 / t_1 - t_0$

The ratio $\Delta d / \Delta t$ is called the average velocity. Average here denotes that this quantity is defined over a period Δt.

ii) Instantaneous velocity is the velocity of an object at a particular moment in time. Conceptually, this can be imagined as the extreme case when Δt is infinitely small.

5. Acceleration represented by a is defined as the rate of change of velocity and the units are m/s^2. Both an average and an instantaneous acceleration can be defined similarly to velocity.

From these definitions we develop the kinematic equations. In the following, subscript i denotes initial and subscript f denotes final values for a time period. Acceleration is assumed to be constant with time.

$$v_f = v_i + at \qquad (1)$$

$$d = v_i t + \frac{1}{2} a t^2 \qquad (2)$$

$$v_f^2 = v_i^2 + 2ad \qquad (3)$$

$$d = \left(\frac{v_i + v_f}{2}\right) t \qquad (4)$$

Example:
Leaving a traffic light a man accelerates at 10 m/s². a) How fast is he going when he has gone 100 m? b) How fast is he going in 4 seconds? C) How far does he travel in 20 seconds.

Solution:
a) Use equation 3. He starts from a stop so v_i=0 and v_f^2=2 x 10m/s² x 100m=2000 m²/s² and v_f=45 m/s.
b) Use equation 1. Initial velocity is again zero so v_f=10m/s² x 4s=40 m/s.
c) Use equation 2. Since initial velocity is again zero, d=1/2 x 10 m/s² x (20s)²=2000 m

Skill 4 Motion in two dimensions (projectile motion, uniform circular motion)

In the previous section, we discussed the relationships between distance, velocity, acceleration and time and the four simple equations that relate these quantities when acceleration is constant (e.g. in cases such as gravity). In two dimensions the same relationships apply, but each dimension must be treated separately.

The most common example of an object moving in two dimensions is a projectile. A projectile is an object upon which the only force acting is gravity. Some examples:
i) An object dropped from rest.
ii) An object thrown vertically upwards at an angle
iii) A canon ball.

Once a projectile has been put in motion (say, by a canon or hand) the only force acting it is gravity, which near the surface of the earth implies it experiences $a=g=9.8 m/s^2$.

This is most easily considered with an example such as the case of a bullet shot horizontally from a standard height at the same moment that a bullet is dropped from exactly the same height. Which will hit the ground first? If we assume wind resistance is negligible, then the acceleration due to gravity is our only acceleration on either bullet and we must conclude that they will hit the ground at the same time. The horizontal motion of the bullet is not affected by the downward acceleration.

Example:
I shoot a projectile at 1000 m/s from a perfectly horizontal barrel exactly 1 m above the ground. How far does it travel before hitting the ground?

Solution:
First figure out how long it takes to hit the ground by analyzing the motion in the vertical direction. In the vertical direction, the initial velocity is zero so we can rearrange kinematic equation 2 from the previous section to give:

$t = \sqrt{\dfrac{2d}{a}}$. Since our displacement is 1 m and $a=g=9.8 m/s^2$, t=0.45 s.

Now use the time to hitting the ground from the previous calculation to calculate how far it will travel horizontally. Here the velocity is 1000m/s and there is no acceleration. So we simple multiply velocity with time to get the distance of 450m.

Motion on an arc can also be considered from the view point of the kinematic equations. As pointed out earlier, displacement, velocity and acceleration are all vector quantities, i.e. they have magnitude (the speed is the magnitude of the velocity vector) and direction. This means that if one drives in a circle at constant speed one still experiences an acceleration that changes the direction. We can define a couple of parameters for objects moving on circular paths and see how they relate to the kinematic equations.

1. Tangential speed: The tangent to a circle or arc is a line that intersects the arc at exactly one point. If you were driving in a circle and instantaneously moved the steering wheel back to straight, the line you would follow would be the tangent to the circle at the point where you moved the wheel. The tangential speed then is the instantaneous magnitude of the velocity vector as one moves around the circle.

2. Tangential acceleration: The tangential acceleration is the component of acceleration that would change the tangential speed and this can be treated as a linear acceleration if one imagines that the circular path is unrolled and made linear.

3. Centripetal acceleration: Centripetal acceleration corresponds to the constant change in the direction of the velocity vector necessary to maintain a circular path. Always acting toward the center of the circle, centripetal acceleration has a magnitude proportional to the tangential speed squared divided by the radius of the path.

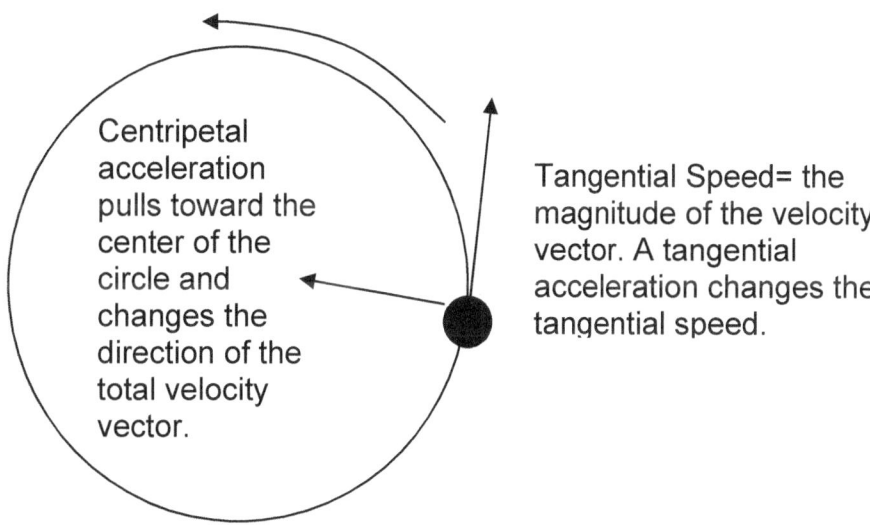

Centripetal acceleration pulls toward the center of the circle and changes the direction of the total velocity vector.

Tangential Speed= the magnitude of the velocity vector. A tangential acceleration changes the tangential speed.

For the forces and equations describing uniform circular motion see section I.10.

Skill 5   Reference frames and relative motion (relative velocity, Galilean relativity)

When we analyze a situation using the laws of physics, we must first consider the perspective from which it is viewed. This is known as the frame of reference. The principles which describe the relationships between different frames of reference are known as relativities. The type of relativity discussed below is known as Galilean or Newtonian relativity and is valid for physical situations in which velocities are relatively low. When velocities approach the speed of light, we must use Einstein's special relativity (see section V.10).

There are two general types of reference frames: inertial and non-inertial

Inertial: These frames translate at a constant vector velocity, meaning the velocity does not change direction or magnitude (i.e., travel in a straight line without acceleration).

Non-inertial: These frames include all other situations in which there is non-constant velocity, such as acceleration or rotation. Galilean relativity does not apply to non-inertial frames, as explained below.

Galilean relativity states that the laws of physics are the same in all inertial frames. That is, these same laws would apply to an experiment performed on the surface of the Earth and an experiment performed in a reference frame moving at constant velocity with respect to the earth. For instance, two baseball players can have the same game of catch either standing on the ground or in a moving bus (so long as the bus's motion has constant direction and magnitude).

It is true, however, that phenomenon will have different appearance depending on our frame of reference. Relative velocity is a useful concept to help us analyze such cases. We can understand relative velocity by again considering the game of catch being played on a bus:

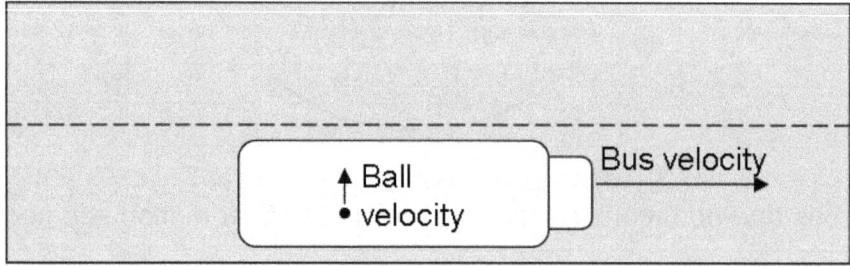

Inside the frame of reference of the bus, the ball travels at the velocity with which it was thrown and straight across the bus (shown by the ball velocity vector above). However, if we use stationary earth as our frame of reference, then the ball is not only moving across the bus, but down the road at the velocity with which the bus is driven. To determine the ball's velocity relative to the earth, then, we must add the ball's velocity relative to the bus and the bus's velocity relative to the earth. This can be performed with simple vector addition.

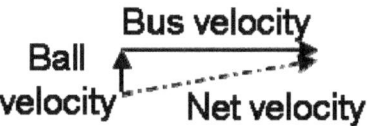

Skill 6   **Force and Newton's laws of motion (Newton's first law, inertia, inertial reference frames, Newton's second law, force and acceleration, addition of forces, balanced versus unbalanced forces, Newton's third law, action-reaction forces, weight and mass)**

Newton's first law of motion: "An object at rest tends to stay at rest and an object in motion tends to stay in motion with the same speed and in the same direction unless acted upon by an unbalanced force". This tendency of an object to continue in its state of rest or motion is known as **inertia**. Note that, at any point in time, most objects have multiple forces acting on them. If the vector addition of all the forces on an object results in a zero net force, then the forces on the object are said to be **balanced**. If the net force on an object is non-zero, an **unbalanced** force is acting on the object.

Prior to Newton's formulation of this law, being at rest was considered the natural state of all objects because at the earth's surface we have the force of gravity working at all times which causes nearly any object put into motion to eventually come to rest. Newton's brilliant leap was to recognize that an unbalanced force changes the motion of a body, whether that body begins at rest or at some non-zero speed.

We experience the consequences of this law everyday. For instance, the first law is why seat belts are necessary to prevent injuries. When a car stops suddenly, say by hitting a road barrier, the driver continues on forward due to inertia until acted upon by a force. The seat belt provides that force and distributes the load across the whole body rather than allowing the driver to fly forward and experience the force against the steering wheel.

Newton's second law of motion: "The acceleration of an object as produced by a net force is directly proportional to the magnitude of the net force, in the same direction as the net force, and inversely proportional to the mass of the object".

In the equation form, it is stated as **F= ma**, force equals mass times acceleration. It is important, again, to remember that this is the net force and that forces are vector quantities. Thus if an object is acted upon by 12 forces that sum to zero, there is no acceleration. Also, this law embodies the idea of inertia as a consequence of mass. For a given force, the resulting acceleration is proportionally smaller for a more massive object because the larger object has more inertia.

The first two laws are generally applied together via the equation **F=ma.** The first law is largely the conceptual foundation for the more specific and quantitative second law. Newton's first law and second law are valid only in **inertial reference frames** (described in previous section).

The **weight** of an object is the result of the gravitational force of the earth acting on its mass. The acceleration due to Earth's gravity on an object is 9.81 m/s². Since force equals mass * acceleration, the magnitude of the gravitational force created by the earth on an object is

$$F_{Gravity} = m_{object} \cdot 9.81 \, m/s^2$$

Example: For the arrangement shown, find the force necessary to overcome the 500 N force pushing to the left and move the truck to the right with an acceleration of 5 m/s².

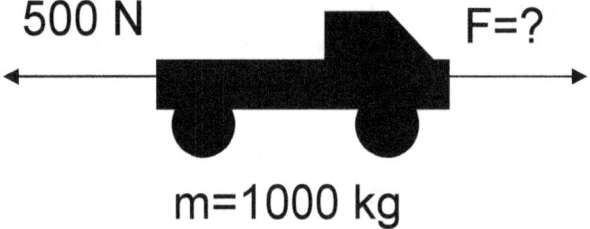

Solution: Since we know the acceleration and mass, we can calculate the net force necessary to move the truck with this acceleration. Assuming that to the right is the positive direction we sum the forces and get
F-500N = 1000kg x 5 m/s². Solving for F, we get 5500 N.

Newton's third law of motion: "For every action, there is an equal and opposite reaction". This statement means that, in every interaction, there is a pair of forces acting on the two interacting objects. The size of the force on the first object equals the size of the force on the second object. The direction of the force on the first object is opposite to the direction of the force on the second object.

1. The propulsion/movement of fish through water: A fish uses its fins to push water backwards. The water pushes back on the fish. Because the force on the fish is unbalanced the fish moves forward.

2. The motion of car: A car's wheels push against the road and the road pushes back. Since the force of the road on the car is unbalanced the car moves forward.

3. Walking: When one pushes backwards on the foot with the muscles of the leg, the floor pushes back on the foot. If the forces of the leg on the foot and the floor on the foot are balanced, the foot will not move and the muscles of the body can move the other leg forward.

Skill 7   Friction (static friction, kinetic friction, rolling friction)

In the real world, whenever an object moves its motion is opposed by a force known as friction. How strong the frictional force is depends on numerous factors such as the roughness of the surfaces (for two objects sliding against each other) or the viscosity of the liquid an object is moving through. Most problems involving the effect of friction on motion deal with sliding friction. This is the type of friction that makes it harder to push a box across cement than across a marble floor.

When you try and push an object from rest, you must overcome the maximum **static friction** force to get it to move. Once the object is in motion, you are working against **kinetic friction** which is smaller than the static friction force previously mentioned. Sliding friction is primarily dependent on two things, the **coefficient of friction (μ)** which is dependent on the roughness of the surfaces involved and the amount of force pushing the two surfaces together. This force is also known as the **normal force (F_n)**, the perpendicular force between two surfaces. When an object is resting on a flat surface, the normal force is pushing opposite to the gravitational force – straight up. When the object is resting on an incline, the normal force is less (because it is only opposing that portion of the gravitational force acting perpendicularly to the object) and its direction is perpendicular to the surface of incline but at an angle from the ground. Therefore, for an object experiencing no external action, the magnitude of the normal force is either equal to or less than the magnitude of the gravitational force (F_g) acting on it. The frictional force (F_f) acts perpendicularly to the normal force, opposing the direction of the object's motion.

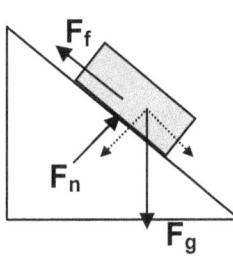

The frictional force is normally directly proportional to the normal force and, unless you are told otherwise, can be calculated as $F_f = \mu F_n$ where μ is either the coefficient of static friction or kinetic friction depending on whether the object starts at rest or in motion. In the first case, the problem is often stated as "how much force does it take to start an object moving" and the frictional force is given by $F_f > \mu_s F_n$ where μ_s is the coefficient of static friction. When questions are of the form "what is the magnitude of the frictional force opposing the motion of this object," the frictional force is given by $F_f = \mu_k F_n$ where μ_k is the coefficient of kinetic friction.

A static frictional force is needed in order to start a ball or a wheel rolling; without this force the object would just slide or spin. **Rolling friction** is the force that resists the rolling motion of an object such as a wheel once it is already in motion. Rolling friction arises from the roughness of the surfaces in contact and from the deformation of the rolling object or surface on which it is rolling. Rolling resistance $F_f = \mu_r F_n$ where μ_r is the coefficient of rolling friction.

There are several important things to remember when solving problems about friction.
1. The frictional force acts in opposition to the direction of motion.
2. The frictional force is proportional to, and acts perpendicular to, the normal force.
3. The normal force is perpendicular to the surface the object is lying on. If there is a force pushing the object against the surface, it will increase the normal force.

Problem:

A woman is pushing an 800N box across the floor. She pushes with a force of 1000 N in the direction indicated in the diagram below. The coefficient of kinetic friction is 0.50. If the box is already moving, what is the force of friction acting on the box?

Solution:
First it is necessary to solve for the normal force.
F_n = 800N + 1000N (sin 30°) = 1300N
Then, since $F_f = \mu F_n$ = 0.5*1300=650N

Skill 8   Equilibrium of forces

An object is said to be in a state of equilibrium when the forces exerted upon it are balanced. That is to say, forces to the left balance the forces exerted to the right, and upward forces are balanced by downward forces. The net force acting on the object is zero and the acceleration is 0 meters per second squared. This does not necessarily mean that the object is at rest. According to Newton's first law of motion, an object at equilibrium is either at rest and remaining at rest (**static equilibrium**), or in motion and continuing in motion with the same speed and direction (**dynamic equilibrium**).

Equilibrium of forces is often used to analyze situations where objects are in static equilibrium. One can determine the weight of an object in static equilibrium or the forces necessary to hold an object at equilibrium. The following are examples of each type of problem.

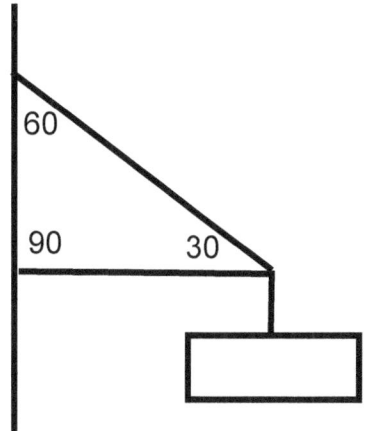

Problem: A sign hangs outside a building supported as shown in the diagram. The sign has a mass of 50 kg. Calculate the tension in the cable.

Solution: Since there is only one upward pulling cable it must balance the weight. The sign exerts a downward force of 490 N. Therefore, the cable pulls upwards with a force of 490 N. It does so at an angle of 30 degrees. To find the total tension in the cable:
$$F_{total} = 490 \text{ N} / \sin 30°$$
$$F_{total} = 980 \text{ N}$$

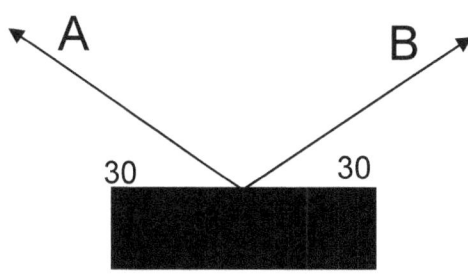

Problem: A block is held in static equilibrium by two cables. Suppose the tension in cables A and B are measured to be 50 Newtons each. The angle formed by each cable with the horizontal is 30 degrees. Calculate the weight of the block.

Solution: We know that the upward pull of the cable must balance the downward force of the weight of the block and the right pulling forces must balance the left pulling forces.

Using trigonometry we know that the y component of each cable can be calculated as:
$$F_y = 50 \text{ N} \sin 30°$$
$$F_y = 25 \text{ N}$$
Since there are two cables supplying an upward force of 25 N each, the overall downward force supplied by the block must be 50 N.

Skill 9 Equilibrium of moments (torques)

For an object to be in equilibrium the forces acting on it must be balanced. This applies to linear as well as rotational forces known as moments or torques. In the two dimensional example below, torque can only be applied in two directions; clockwise and counter clockwise. The convention is that positive rotation is counter clockwise and negative is clockwise. For the object to be in equilibrium, the sum of the applied torques must be zero, in addition to the sum of all forces being zero.

Let us consider a horizontal bar at equilibrium so that the bar experiences neither rotation nor translation. We can define rotation by choosing any point along the bar and labeling it A for axis.

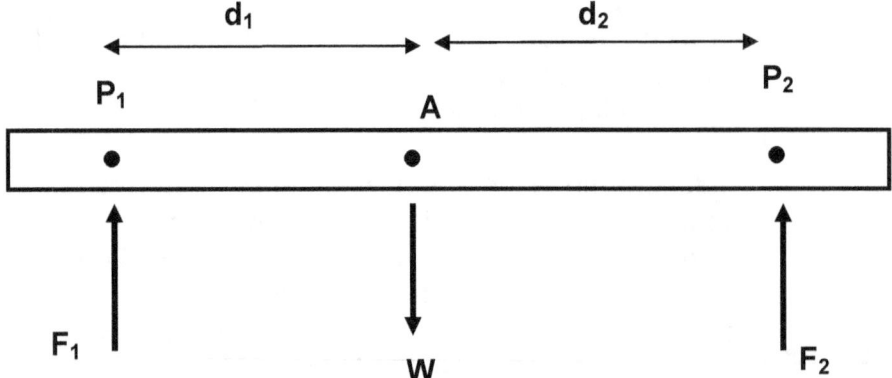

The bar experiences two point forces at either end, labeled F_1 and F_2. The torque applied to the bar by each of these forces is given by multiplying the force by the moment arm, the distance between the point where the force is applied and the axis. In the case of F_1 the torque is as follows (negative since rotation is clockwise):

$$\tau_1 = -F_1 d_1$$

The torque applied by F2 is given by (positive since rotation is counterclockwise):

$$\tau_2 = F_2 d_2$$

The other force acting on the bar is the force of gravity, or the weight of the bar. This is not a point force, but rather acts at all points along the bar. However, we can consider that the weight acts in the center of the bar, at a point called the center of mass. In the case of this example, we are taking the axis to be located at the center of mass. Since the axis is located at the center of mass, the torque exerted on the bar due to its weight is zero.

Suppose F_1=2 N, d_1=0.4 m, and d_2=0.5 m. Let us calculate F_2. We will use our knowledge that the sum of all torques must equal zero when that object is at equilibrium.

$$\tau_1 + \tau_2 = 0$$

$$-F_1 d_1 + F_2 d_2 = 0$$
$$(-2 \text{ N} \times 0.4 \text{ m}) + (F_2 \times 0.5 \text{ m}) = 0$$
$$-0.8 + 0.5 F_2 = 0$$
$$F_2 = 1.6 \text{ N}$$

It is also possible to calculate the weight of the bar since we know that the sum of all forces must be zero. Since F1 and F2 act up but weight acts down we have:

$$2\,N + 1.6\,N - W = 0$$

$$W = 3.6\,N$$

Skill 10   Dynamics of uniform circular motion

Uniform circular motion describes the motion of an object as it moves in a circular path at constant speed. There are many everyday examples of this behavior though we may not recognize them if the object does not complete a full circle. For example, a car rounding a curve (that is an arc of a circle) often exhibits uniform circular motion.

The following diagram and variable definitions will help us to analyze uniform circular motion.

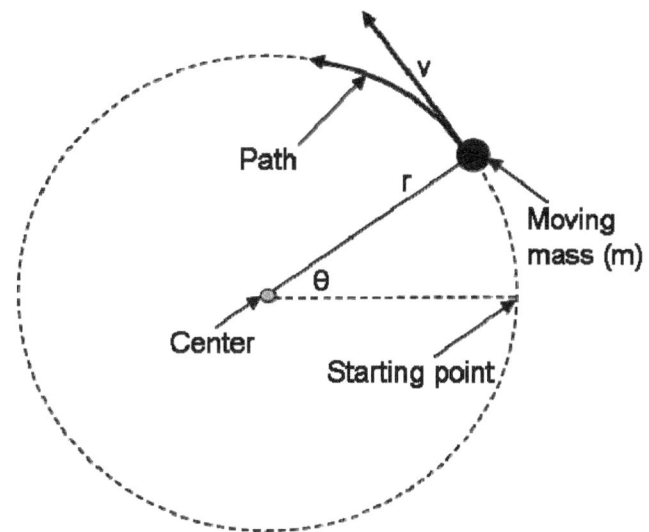

Above we see that the mass is traveling a path with constant radius (r) from some center point (x_0, y_0). By defining a variable (θ) that is a function of time (t) and is the angle between the mass's present position and original position on the circular path, we can write the following equations for the mass's position in a Cartesian plane.

$$x = r\cos(\theta) + x_0$$
$$y = r\sin(\theta) + y_0$$

Next observe that, because we are discussing uniform circular motion, the *magnitude* of the mass's velocity (v) is constant. However, the velocity's direction is always tangent to the circle and so always changing. We know that a changing velocity means that the mass must have a positive acceleration. This acceleration is directed toward the center of the circular path and is always perpendicular to the velocity, as shown below:

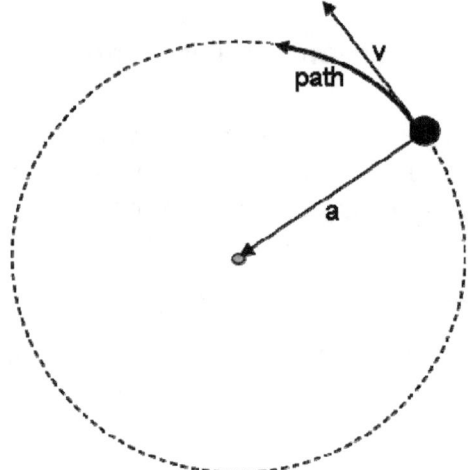

This is known as centripetal acceleration and is mathematically expressed as:

$$a = \frac{v^2}{r} = \frac{4\pi^2 r}{t^2}$$

where t is the period of the motion or the time taken for the mass to travel once around the circle. The force (F) experienced by the mass (m) is known as centripetal force and is always directed towards the center of the circular path. It has constant magnitude given by the following equation:

$$F = ma = m\frac{v^2}{r}$$

Skill 11 **Work, energy, and power (relationship between work and kinetic energy, work done by a variable force)**

In physics, work done by a constant force is defined as force times distance $W = F \cdot s$. Work is a scalar quantity, it does not have direction, and it is usually measured in Joules ($N \cdot m$). It is important to remember, when doing calculations about work, that the only part of the force that contributes to the work is the part that acts in the direction of the displacement. Therefore, sometimes the equation is written as $W = F \cdot s \cos\theta$, where θ is the angle between the force and the displacement.

When the force applied to an object varies over the distance moved, the total work done in moving from a point x_1 to a point x_2 is given by

$$W = \int_{x_1}^{x_2} F_x \, dx$$

where F_x is the force applied to the object, in the direction of movement, at any point x.

Problem: A man uses a constant 6N force to pull a 10kg block, as shown below, over a distance of 3 m. How much work did he do?

Solution:

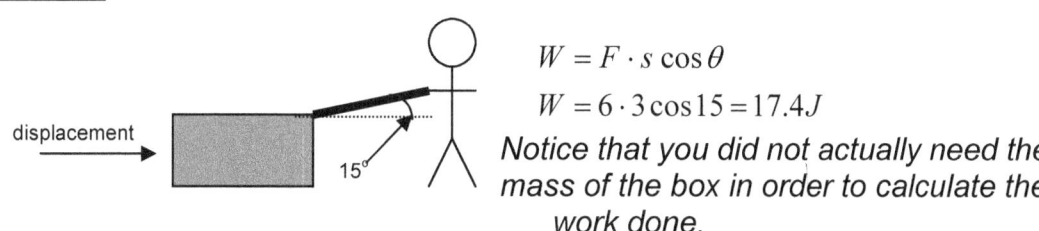

$W = F \cdot s \cos \theta$
$W = 6 \cdot 3 \cos 15 = 17.4 J$

Notice that you did not actually need the mass of the box in order to calculate the work done.

Power is defined in relationship to work. It is the rate at which work is done, or, in other words, the amount of work done in a certain period of time: $P = W/t$. There are many different units for power, but the one most commonly seen in physics problems is the Watt which is measured in Joules per second. Another commonly discussed unit of power is horsepower, and 1hp=746 W.

Problem: A woman standing in her 4th story apartment raises a 10kg box of groceries from the ground using a rope. She is pulling at a constant rate, and it takes her 5 seconds to raise the box one meter. How much power is she using to raise the box?

Solution:

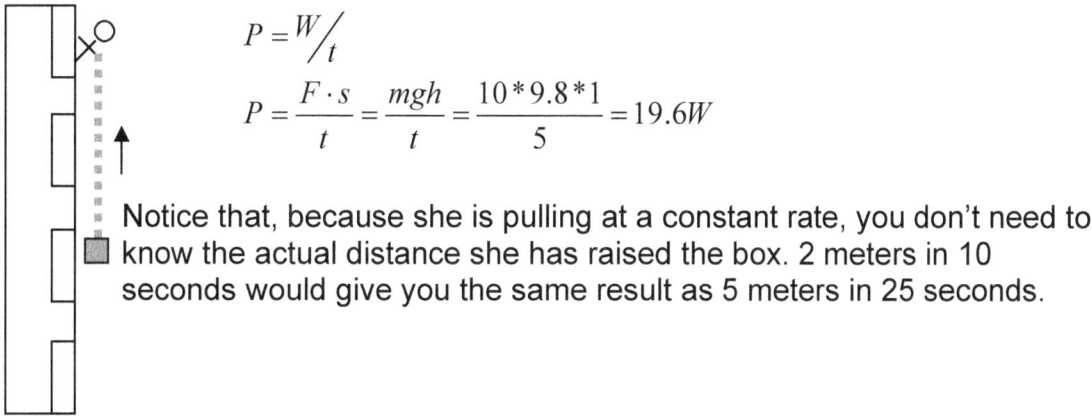

$P = W/t$

$P = \dfrac{F \cdot s}{t} = \dfrac{mgh}{t} = \dfrac{10 * 9.8 * 1}{5} = 19.6W$

Notice that, because she is pulling at a constant rate, you don't need to know the actual distance she has raised the box. 2 meters in 10 seconds would give you the same result as 5 meters in 25 seconds.

PHYSICS

Energy is also defined, in relation to work, as the ability of an object to do work. As such, it is measured in the same units as work, usually Joules. Most problems relating work to energy are looking at two specific kinds of energy. The first, kinetic energy, is the energy of motion. The heavier an object is and the faster it is going, the more energy it has resulting in a greater capacity for work. The equation for kinetic energy is : $KE = \frac{1}{2}mv^2$.

Problem:

A 1500 kg car is moving at 60m/s down a highway when it crashes into a 3000kg truck. In the moment before impact, how much kinetic energy does the car have?

Solution:

$$KE = \tfrac{1}{2}mv^2 = \tfrac{1}{2} \cdot 1500 \cdot 60^2 = 2.7 \times 10^6 J$$

The other form of energy frequently discussed in relationship to work is gravitational potential energy, or potential energy, the energy of position. Potential energy is calculated as $PE = mgh$ where h is the distance the object is capable of falling.

Problem:

Which has more potential energy, a 2 kg box held 5 m above the ground or a 10 kg box held 1 m above the ground?

Solution:

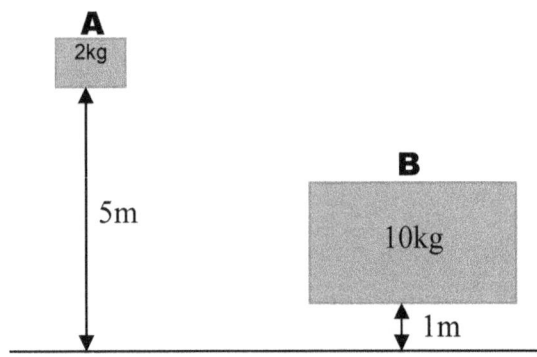

$$PE_A = mgh = 2 \cdot g \cdot 5 = 10g$$
$$PE_B = mgh = 10 \cdot g \cdot 1 = 10g$$
$$PE_A = PE_B$$

Skill 12   Conservation of energy (potential energy, conservative and nonconservative forces)

The principle of conservation of energy states that an isolated system maintains a constant total amount of energy despite the fact that the energy may change forms. To put it another way, energy cannot be created or destroyed but can be changed from one form to another. For example, friction can turn kinetic energy into thermal energy. Other forms of energy include electrical energy, chemical energy, and mechanical energy.

A **conservative** force is one that conserves mechanical energy (kinetic + potential energy), i.e. there is no change in mechanical energy when a conservative force acts on an object. Consider a mass on a spring on a frictionless surface. This is a closed loop system. If conservative forces alone act on the mass during each cycle, the velocity of the mass at the beginning and the end of the cycle must be the same for the mechanical energy to have been conserved. In this way, the force has done no work. At any point in the cycle of motion, the total mechanical energy of the system remains constant even though the energy moves back and forth between kinetic and potential forms. If work is done on the mass, then the forces acting on the mass are **nonconservative**. In a real system there will be some dissipative forces that will convert some of the mechanical energy to thermal energy. Conservative forces are independent of path the object takes, while nonconservative forces are path dependent.

Gravity is a conservative force. This can be illustrated by imagining an object tossed into the air. On the upward journey the work done by gravity is the negative product of mass, acceleration, and height. On the downward journey, the work done by gravity is the positive value of this amount. Thus for the total loop the work is zero.

Friction is a nonconservative force. If a box is pushed along a rough surface from one side of the room to the other and back, friction opposes the movement in both directions; so the work done by friction cannot be equal to zero. This example also helps illustrate how nonconservative forces are path dependent. More work is done by friction if the path is tortuous rather than straight, even if the start and end points are the same. Let's try an example with a small box of mass 5 kg. The box moves in a circle 2 meters in diameter. The coefficient of kinetic friction between the box and the surface it rests on is 0.2. How much work is done by friction during one revolution?

The force exerted by friction is calculated by

$$F_k = \mu_k F_n = (0.2)(5 \text{ kg})(9.8 \text{ m/s}^2) = 9.8 \text{ N}$$

The force opposes the movement of the box during the entire distance of one revolution, or approximately 6.3 meters ($2\pi r$).

The total work done by friction is

W=F x cos θ= (9.8 N) (6.3 m) (cos 180)= -61.7 Joules

As expected, the work is not zero since friction is not a conservative force. Since it does negative work on an object, it reduces the mechanical energy of the object and is a **dissipative** force.

Skill 13   Simple harmonic motion and oscillations (Hooke's law, graphical and mathematical representations, energy considerations, pendulums, springs)

Harmonic motion or harmonic oscillation is seen in any system that follows **Hooke's Law**. Hooke's law simply predicts the behavior of certain bodies as they return to equilibrium following a displacement. It is given by the following equation:

$$F = -kx$$

Where F=restoring force
x=displacement
k=a positive constant

From this equation we can see that the harmonic motion is neither damped nor driven. Harmonic motion is observed as sinusoidal oscillations about an equilibrium point. Both the amplitude and the frequency are constant. Further, the amplitude is always positive and is a function of the original force that disrupted the equilibrium. If an object's oscillation is governed solely by Hooke's law, it is a simple harmonic oscillator. Below are examples of simple harmonic oscillators:

Pendula: A pendulum is a mass on the end of a rigid rod or a string. An initial push will cause the pendulum to swing back and forth. This motion will be harmonic as long as the pendulum moves through an angle of less than 15°.

Masses connected to springs: A spring is simply the familiar helical coil of metal that is used to store mechanical energy. In a typical system, one end of a spring is attached to a mass and the other to a solid surface (a wall, ceiling, etc). If the spring is then stretched or compressed (i.e., removed from equilibrium) it will oscillate harmonically.

Vibrating strings: A string or rope tied tightly at both ends will oscillate harmonically when it is struck or plucked. This is often the mechanism used to generate sound in string-based instruments such as guitars and pianos.

Several simple harmonic oscillations maybe superimposed to create complex harmonic motion. The best-known example of complex harmonic motion is a musical chord.

View an animation of harmonic oscillation here:
http://en.wikipedia.org/wiki/Image:Simple_harmonic_motion_animation.gif

The displacement of a simple harmonic oscillator varies sinusoidally with time and is given by

$$x = A\cos(\omega t + \delta)$$

where A is the maximum displacement or amplitude, ω is the angular frequency and δ is the phase constant. We can see from the equation that the displacement goes through a full cycle at time intervals given by the period $T = 2\pi/\omega$. The figure below displays a graphical representation of the displacement of a simple harmonic oscillator.

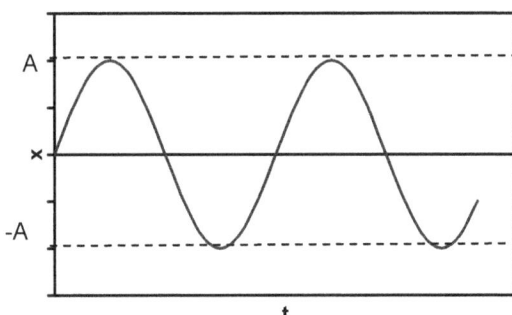

In the absence of dissipative forces, the total energy of a simple harmonic oscillator remains constant; however, the proportion of kinetic to potential energy varies. At x = A and x = -A all of the energy is potential. At x = 0 all of the energy is kinetic.

Skill 14 Linear momentum and impulse (momentum-impulse relationship, conservation of linear momentum; elastic and inelastic collisions)

The law of **conservation of linear momentum** states that the total momentum of an *isolated system* (not affected by external forces and not having internal dissipative forces) always remains the same. For instance, in any collision between two objects in an isolated system, the total momentum of the two objects after the collision will be the same as the total momentum of the two objects before the collision. In other words, any momentum lost by one of the objects is gained by the other.

A collision may be **elastic** or **inelastic**. In a totally elastic collision, the kinetic energy is conserved along with the momentum. In a totally inelastic collision, on the other hand, the kinetic energy associated with the center of mass remains unchanged but the kinetic energy relative to the center of mass is lost. An example of a totally inelastic collision is one in which the bodies stick to each other and move together after the collision. Most collisions are neither perfectly elastic nor perfectly inelastic and only a portion of the kinetic energy relative to the center of mass is lost.

Imagine two carts rolling towards each other as in the diagram below

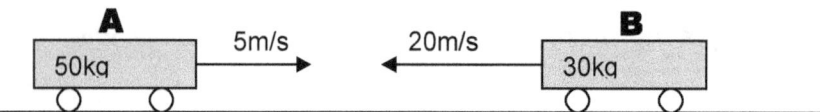

Before the collision, cart **A** has 250 kg m/s of momentum, and cart **B** has –600 kg m/s of momentum. In other words, the system has a total momentum of –350 kg m/s of momentum.

After the inelastic collision, the two cards stick to each other, and continue moving. How do we determine how fast, and in what direction, they go?

We know that the new mass of the cart is 80kg, and that the total momentum of the system is –350 kg m/s. Therefore, the velocity of the two carts stuck together must be $\frac{-350}{80} = -4.375 \, m/s$

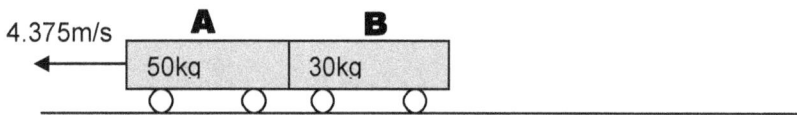

Conservation of momentum works the same way in two dimensions, the only change is that you need to use vector math to determine the total momentum and any changes, instead of simple addition.

Imagine a pool table like the one below. Both balls are 0.5 kg in mass.

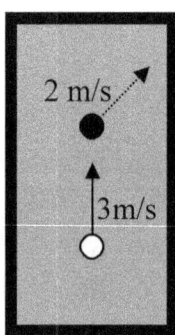

Before the collision, the white ball is moving with the velocity indicated by the solid line and the black ball is at rest.
After the collision the black ball is moving with the velocity indicated by the dashed line (a 135° angle from the direction of the white ball).

With what speed, and in what direction, is the white ball moving after the collision?

$p_{white/before} = .5 \cdot (0,3) = (0,1.5)$ $p_{black/before} = 0$ $p_{total/before} = (0,1.5)$

$p_{black/after} = .5 \cdot (2\cos 45, 2\sin 45) = (0.71, 0.71)$

$p_{white/after} = (-0.71, 0.79)$

i.e. the white ball has a velocity of $v = \sqrt{(-.71)^2 + (0.79)^2} = 1.06 \, m/s$

and is moving at an angle of $\theta = \tan^{-1}\left(\frac{0.79}{-0.71}\right) = -48°$ from the horizontal

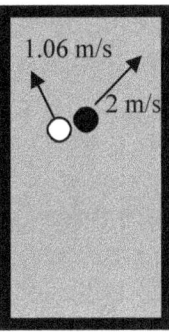

The **impulse-momentum theorem** states that any impulse acting on a system changes the momentum of that system. When considering the impulse-momentum theorem, there are several factors that need to be taken into account. The first factor is that momentum is a vector quantity $p = m \cdot v$. It has both magnitude and direction. Therefore, any action that causes either the speed or the direction of an object to change causes a change in its momentum. An impulse is defined as a force acting over a period of time (integral of force over time), and any impulse acting on the system is equivalent to a change in its momentum, as you can see from the equations below:

$$F = m \cdot a \rightarrow F = m \cdot \frac{\Delta v}{t} \rightarrow F \cdot t = m \cdot \Delta v$$

i.e. Forces acting over time cause a change in momentum.

Sample Problems:

1. A 1 kg ball is rolled towards a wall at 4 m/s. It hits the wall, and bounces back off the wall at 3 m/s.

   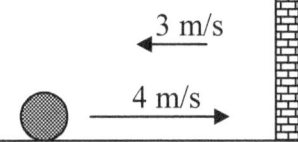

   a. What is the change in velocity?

   The velocity goes from +4m/s to –3m/s, a net change of -7m/s.

At what point does the impulse occur?

   The impulse occurs when the ball hits the wall.

2. A 30kg woman is in a car accident. She was driving at 50m/s when she had to hit the brakes to avoid hitting the car in front of her.
   a. The automatic tensioning device in her seatbelt slows her down to a stop over a period of one half second. How much force does it apply?
   $$F = m \cdot \frac{\Delta v}{t} \rightarrow F = 30 \cdot \frac{50}{.5} = 3000N$$
   b. If she hadn't been wearing a seatbelt, the windshield would have stopped her in .001 seconds. How much force would have been applied there?
   $$F = m \cdot \frac{\Delta v}{t} \rightarrow F = 30 \cdot \frac{50}{.001} = 1500000N$$

Skill 15   **Rigid body motion (angular velocity and angular acceleration, angular momentum, moment of inertia, torque, and center of mass, conservation of angular momentum, rotational kinetic energy)**

Linear motion is measured in rectangular coordinates. Rotational motion is measured differently, in terms of the angle of displacement. There are three common ways to measure rotational displacement; degrees, revolutions, and radians. Degrees and revolutions have an easy to understand relationship, one revolution is 360°. Radians are slightly less well known and are defined as

$\dfrac{arc\ length}{radius}$. Therefore 360°=2π radians and 1 radian = 57.3°.

The major concepts of linear motion are duplicated in rotational motion with linear displacement replaced by **angular displacement**.

Angular velocity ω = rate of change of angular displacement.
Angular acceleration α = rate of change of angular velocity.

Also, the linear velocity v of a rolling object can be written as $v = r\omega$ and the linear acceleration as $a = r\alpha$.

One important difference in the equations relates to the use of mass in rotational systems. In rotational problems, not only is the mass of an object important but also its location. In order to include the spatial distribution of the mass of the object, a term called **moment of inertia** is used, $I = m_1 r_1^2 + m_2 r_2^2 + \cdots + m_n r_n^2$. The moment of inertia is always defined with respect to a particular axis of rotation.

Example:

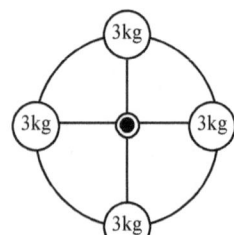

If the radius of the wheel on the left is 0.75m, what is its moment of inertia about an axis running through its center perpendicular to the plane of the wheel?

$$I = 3 \cdot 0.75^2 + 3 \cdot 0.75^2 + 3 \cdot 0.75^2 + 3 \cdot 0.75^2 = 6.75$$

Note: $I_{Sphere} = \dfrac{2}{5}mr^2$, $I_{Hoop/Ring} = mr^2$, $I_{disk} = \dfrac{1}{2}mr^2$

The rotational analog of Newton's second law of motion is given in terms of **torque** τ, moment of inertia I, and angular acceleration α:

$$\tau = I\alpha$$

where the torque τ is the rotational force on the body. In simple terms, the torque τ produced by a force F acting at a distance r from the point of rotation is given by the product of r and the component of the force that is perpendicular to the line joining the point of rotation to the point of action of the force.

A concept related to the moment of inertia is the **radius of gyration** (k), which is the average distance of the mass of an object from its axis of rotation, i.e., the distance from the axis where a point mass m would have the same moment of inertia.

$k_{Sphere} = \sqrt{\frac{2}{5}}r$, $k_{Hoop/Ring} = r$, $k_{disk} = \frac{r}{\sqrt{2}}$. As you can see $I = mk^2$

This is analogous to the concept of center of mass, the point where an equivalent mass of infinitely small size would be located, in the case of linear motion.

Angular momentum (L), and **rotational kinetic energy (KE_r),** are therefore defined as follows: $L = I\omega$, $KE_r = \frac{1}{2}I\omega^2$

As with all systems, energy is conserved unless the system is acted on by an external force. This can be used to solve problems such as the one below.

Example:

A uniform ball of radius *r* and mass *m* starts from rest and rolls down a frictionless incline of height *h*. When the ball reaches the ground, how fast is it going?

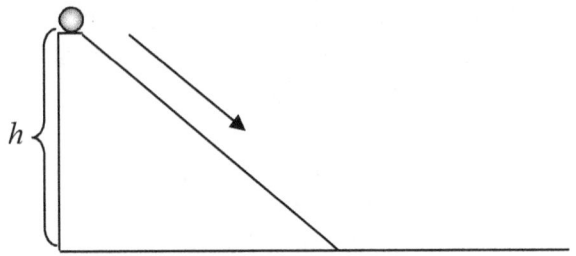

$$PE_{initial} + KE_{rotational/initial} + KE_{linear/initial} = PE_{final} + KE_{rotational/final} + KE_{linear/final}$$

$$mgh + 0 + 0 = 0 + \frac{1}{2}I\omega_{final}^2 + \frac{1}{2}mv_{final}^2 \rightarrow mgh = \frac{1}{2}\cdot\frac{2}{5}mr^2\omega_{final}^2 + \frac{1}{2}mv_{final}^2$$

$$mgh = \frac{1}{5}mr^2(\frac{v_{final}}{r})^2 + \frac{1}{2}mv_{final}^2 \rightarrow mgh = \frac{1}{5}mv_{final}^2 + \frac{1}{2}mv_{final}^2$$

$$gh = \frac{7}{10}v_{final}^2 \rightarrow v_{final} = \sqrt{\frac{10}{7}gh}$$

Similarly, unless a net torque acts on a system, the angular momentum remains constant in both magnitude and direction. This can be used to solve many different types of problems including ones involving satellite motion.

Example:
A planet of mass *m* is circling a star in an orbit like the one below. If its velocity at point A is 60,000m/s, and $r_B = 8\, r_A$, what is its velocity at point B?

$$I_B\omega_B = I_A\omega_A$$
$$mr_B^2\omega_B = mr_A^2\omega_A$$
$$r_B^2\omega_B = r_A^2\omega_A$$
$$r_B^2\frac{v_B}{r_B} = r_A^2\frac{v_A}{r_A}$$
$$r_B v_B = r_A v_A$$
$$8r_A v_B = r_A v_A$$
$$v_B = \frac{v_A}{8} = 7500 m/s$$

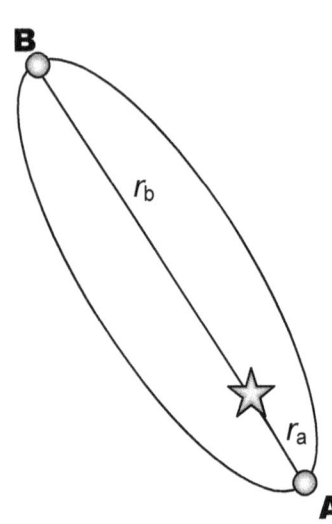

Skill 16   Mass-energy relationships (conservation of mass-energy)

Mass-energy equivalence is the principle that mass is a form of energy. Even when a body of mass m is at rest and has no kinetic energy, it has a **rest energy** E given by Einstein's famous equation $E = mc^2$, a result derived from his special theory of relativity. The total relativistic energy of a body of mass m moving at a velocity v is given by $E = \dfrac{mc^2}{\sqrt{1 - v^2/c^2}}$. We can see that when the body is at rest $v=0$ and rest energy $E = mc^2$.

One consequence of mass-energy equivalence is that the principles of conservation of mass and conservation of energy are combined into the **conservation of mass and energy** as a whole. The other corollary is that mass may be transformed into other types of energy and vice versa.

A notable feature of the transformation of mass into energy is the fact that the conversion factor c^2 is extremely large. Thus a very small mass may yield a stupendous amount of energy, a significant factor in the development of the atomic bomb.

Problem: What is the quantity of energy contained in a 1g mass?
Solution: $E = mc^2 \approx 10^{-3} \times (3 \times 10^8)^2 = 9 \times 10^{13} J$

Examples of the conservation of mass-energy are given in sections V.2 and V.7.

Skill 17   Newton's law of universal gravitation and orbital motion (motion of satellites)

Newton's universal law of gravitation states that any two objects experience a force between them as the result of their masses. Specifically, the force between two masses m_1 and m_2 can be summarized as

$$F = G\frac{m_1 m_2}{r^2}$$

where G is the gravitational constant ($G = 6.672 \times 10^{-11} Nm^2/kg^2$), and r is the distance between the two objects.

Important things to remember:
1. The gravitational force is proportional to the masses of the two objects, but *inversely* proportional to the *square of the distance* between the two objects.
2. When calculating the effects of the acceleration due to gravity for an object above the earth's surface, the distance above the surface is ignored because it is inconsequential compared to the radius of the earth. The constant figure of 9.81 m/s² is used instead.

Problem: Two identical 4 kg balls are floating in space, 2 meters apart. What is the magnitude of the gravitational force they exert on each other?

Solution:

$$F = G\frac{m_1 m_2}{r^2} = G\frac{4 \times 4}{2^2} = 4G = 2.67 \times 10^{-10}\, N$$

For a satellite of mass m in orbit around the earth (mass M), the gravitational attraction of the earth provides the centripetal force that keeps the satellite in motion:

$$\frac{GMm}{r^2} = \frac{mv^2}{r} = mr\omega^2 = mr\left(\frac{2\pi}{T}\right)^2$$

Thus the period T of rotation of the satellite may be obtained from the equation

$$\frac{T^2}{r^3} = \frac{4\pi^2}{GM}$$

(Compare this with Kepler's third law in the next section.)

Skill 18 Kepler's laws (law of orbits (first law), law of areas (second law), law of periods (third law))

Johannes Kepler was a German mathematician who studied the astronomical observations made by Tyco Brahe. He derived the following three laws of planetary motion. Kepler's laws also predict the motion of comets.

First law

This law describes the shape of planetary orbits. Specifically, the orbit of a planet is an ellipse that has the sun at one of the foci. Such an orbit looks like this:

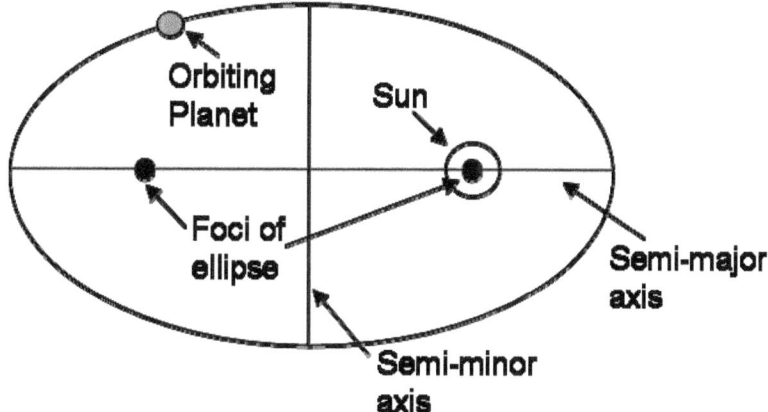

To analyze this situation mathematically, remember that the semi-major axis is denoted a, the semi-minor axis denoted b, and the general equation for an ellipse in polar coordinates is:

$$r = \frac{l}{1 + e\cos\theta}$$

Where r=radial coordinate
θ=angular coordinate
l= semi-latus rectum (l=b²/a)
e=eccentricity (for an ellipse, $e = \sqrt{1 - \frac{b^2}{a^2}}$)

Thus, we can also determine the planet's maximum and minimum distance from the sun.

The point at which the planet is closest to the sun is known as the perihelion and occurs when θ=0:

$$r_{min} = \frac{l}{1+e}$$

The point at which the planet is farthest from the sun is known as the aphelion and occurs when θ=180°:

$$r_{max} = \frac{l}{1-e}$$

Second Law

The second law pertains to the relative speed of a planet as it orbits. This law says that a line joining the planet and the Sun sweeps out equal areas in equal intervals of time. In the diagram below, the two shaded areas demonstrate equal areas.

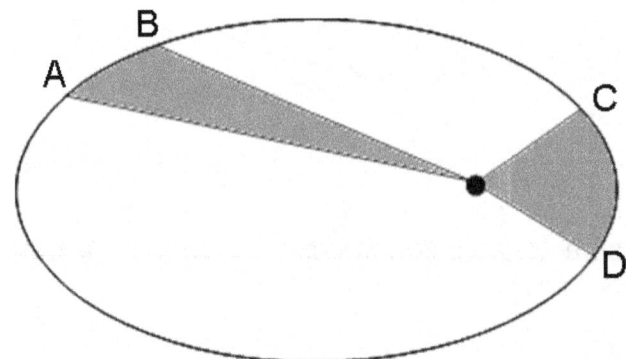

By Kepler's second law, we know that the planet will take the same amount of time to move between points A and B and between points C and D. Note that this means that the speed of the planet is inversely proportional to its distance from the sun (i.e., the plant moves fastest when it is closest to the sun). You can view an animation of this changing speed here:

http://home.cvc.org/science/kepler.gif

Kepler was only able to demonstrate the existence of this phenomenon but we now know that it is an effect of the Sun's gravity. The gravity of the Sun pulls the planet toward it thereby accelerating the planet as it nears. Using the first two laws together, Kepler was able to calculate a planet's position from the time elapsed since the perihelion.

Third law

The third law is also known as the harmonic law and it relates the size of a planet's orbital to the time needed to complete it. It states that the square of a planet's period is proportional to the cube of its mean distances from the Sun (this mean distance can be shown to be equal to the semi-major axis). So, we can state the third law as:

$$P^2 \propto a^3$$

where P=planet's orbital period (length of time needed to complete one orbit)
a=semi-major axis of orbit

Furthermore, for two planets A and B:

$$P_A^2 / P_B^2 = a_A^3 / a_B^3$$

The units for period and semi-major axis have been defined such that $P^2 a^{-3}=1$ for all planets in our solar system. These units are sidereal years (yr) and astronomical units (AU). Sample values are given in the table below. Note that in each case $P^2 \sim a^3$

Planet	P (yr)	a (AU)	P^2	a^3
Venus	0.62	0.72	0.39	0.37
Earth	1.0	1.0	1.0	1.0
Jupiter	11.9	5.20	142	141

Skill 19 **Fluids (density and pressure, ideal fluids at rest, Pascal's law, Archimedes' principle and buoyant forces)**

The weight of a column of fluid creates hydrostatic pressure. Common situations in which we might analyze hydrostatic pressure include tanks of fluid, a swimming pool, or the ocean. Also, atmospheric pressure is an example of hydrostatic pressure.

Because hydrostatic pressure results from the force of gravity interacting with the mass of the fluid or gas, for an incompressible fluid it is governed by the following equation:

$$P = \rho g h$$

where P=hydrostatic pressure
ρ=density of the fluid
g=acceleration of gravity
h=height of the fluid column

Example: How much pressure is exerted by the water at the bottom of a 5 meter swimming pool filled with water?

Solution: We simply use the equation from above, recalling that the acceleration due to gravity is 9.8m/s² and the density of water is 1000 kg/m³.

$$P = \rho g h = 1000 \frac{kg}{m^3} \times 9.8 \frac{m}{s^2} \times 5m = 49,000 Pa = 49 kPa$$

PHYSICS

According to **Pascal's principle**, when pressure is applied to an enclosed fluid, it is transmitted undiminished to all parts of the fluid. For instance, if an additional pressure P_0 is applied to the top surface of a column of liquid of height h as described above, the pressure at the bottom of the liquid will increase by P_0 and will be given by $P = P_0 + \rho g h$. This principle is used in devices such as a **hydraulic lift** (shown below) which consists of two fluid-filled cylinders, one narrow and one wide, connected at the bottom. Pressure P (force = P X A1) applied on the surface of the fluid in the narrow cylinder is transmitted undiminished to the wider cylinder resulting in a larger net force (P X A2) transmitted through its surface. Thus a relatively small force is used to lift a heavy object. This does not violate the conservation of energy since the small force has to be applied through a large distance to move the heavy object a small distance.

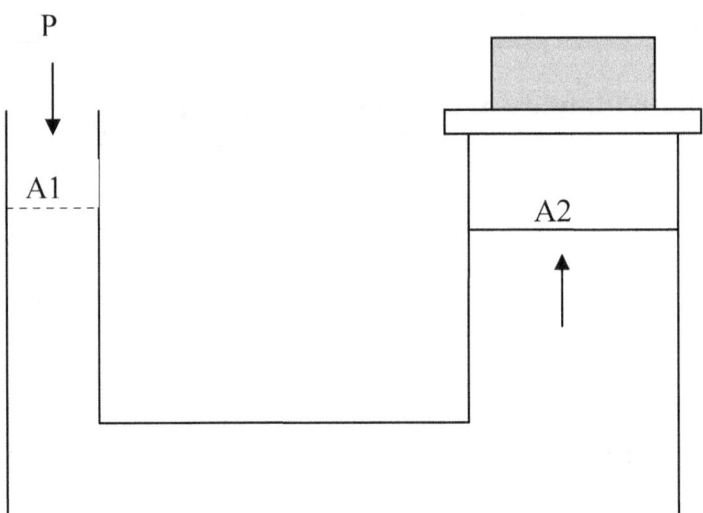

Archimedes' Principle states that, for an object in a fluid, "the upthrust is equal to the weight of the displaced fluid" and the weight of displaced fluid is directly proportional to the volume of displaced fluid. The second part of his discovery is useful when we want to, for instance, determine the volume of an oddly shaped object. We determine its volume by immersing it in a graduated cylinder and measuring how much fluid is displaced. We explore his first observation in more depth below.

Today, we call Archimedes' "upthrust" **buoyancy**. Buoyancy is the force produced by a fluid on a fully or partially immersed object. The buoyant force (F$_{buoyant}$) is found using the following equation:

$$F_{bouyant} = \rho V g$$

where ρ=density of the fluid
V=volume of fluid displaced by the submerged object
g=the acceleration of gravity

Notice that the buoyant force opposes the force of gravity. For an immersed object, a gravitational force (equal to the object's mass times the acceleration of gravity) pulls it downward, while a buoyant force (equal to the weight of the displaced fluid) pushes it upward.

Also note that, from this principle, we can predict *whether* an object will sink or float in a given liquid. We can simply compare the density of the material from which the object is made to that of the liquid. If the material has a lower density, it will float; if it has a higher density it will sink. Finally, if an object has a density equal to that of the liquid, it will neither sink nor float.

Example: Will gold (ρ=19.3 g/cm³) float in water?

Solution: We must compare the density of gold with that of water, which is 1 g/cm³.

$$\rho_{gold} > \rho_{water}$$

So, gold will sink in water.

Example: Imagine a 1 m³ cube of oak (530 kg/m³) floating in water. What is the buoyant force on the cube and how far up the sides of the cube will the water be?

Solution: Since the cube is floating, it has displaced enough water so that the buoyant force is equal to the force of gravity. Thus the buoyant force on the cube is equal to its weight 1X530X9.8 N = 5194 N.

To determine where the cube sits in the water, we simply the find the ratio of the wood's density to that of the water:

$$\frac{\rho_{oak}}{\rho_{water}} = \frac{530 \, kg/m^3}{1000 \, kg/m^3} = 0.53$$

Thus, 53% of the cube will be submerged. Since the edges of the cube must be 1m each, the top 0.47m of the cube will appear above the water.

Skill 20   Ideal fluids in motion (Bernoulli's principle, streamlines, equation of continuity)

The study of moving fluids is contained within fluid mechanics which is itself a component of continuum mechanics. Some of the most important applications of fluid mechanics involve liquids and gases moving in tubes and pipes.

Fluid flow may be laminar or turbulent. One cannot predict the exact path a fluid particle will follow in turbulent or erratic flow. Laminar flow, however, is smooth and each fluid particle follows a continuous path. Lines, know as **streamlines,** are drawn to show the path of a laminar fluid. Streamlines never cross one another and higher fluid velocity is depicted by drawing streamlines closer together.

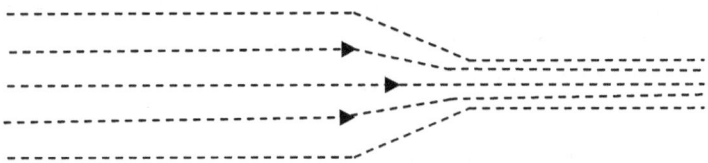

To understand the movement of laminar fluids, one of the first quantities we must define is volumetric flow rate which may have units of gallons per min (gpm), liters/s, cubic feet per min (cfm), gpf, or m³/s:

$$Q = Av\cos\theta$$

> Where Q=volumetric flow rate
> A=cross sectional area of the pipe
> v=fluid velocity
> θ=the angle between the direction of the fluid flow and a vector normal to A

Note that in situations in which the fluid velocity is perpendicular to the cross sectional area, this equation is simply:

$$Q = Av$$

It is also convenient to sometimes discuss mass flow rate ($\dot{m}$), which we can easily find using the density (ρ) of the fluid:

$$\dot{m} = \rho vA = \rho Q$$

Usually, we make an assumption that the fluid is incompressible, that is, the density is constant. Like many commonly used simplifications, this assumption is largely and typically correct though real fluids are, of course, compressible to varying extents. When we do assume that density is constant, we can use conservation of mass to determine that when a pipe is expanded or restricted, the mass flow rate will remain the same. Let's see how this pertains to an example:

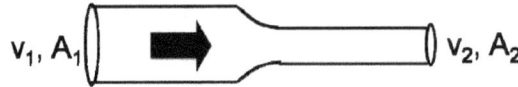

Given conservation of mass, it must be true that:

$$v_1 A_1 = v_2 A_2$$

This is known as the **equation of continuity**. Note that this means the fluid will flow faster in the narrower portions of the pipe and more slowly in the wider regions. An everyday example of this principle is seen when one holds their thumb over the nozzle of a garden hose; the cross sectional area is reduced and so the water flows more quickly.

Much of what we know about fluid flow today was originally discovered by Daniel Bernoulli. His most famous discovery is known as Bernoulli's Principle which states that, if no work is performed on a fluid or gas, an increase in velocity will be accompanied by a decrease in pressure. The mathematical statement of the Bernoulli's Principle for incompressible flow is:

$$\frac{v^2}{2} + gh + \frac{p}{\rho} = \text{constant}$$

where v = fluid velocity
g = acceleration due to gravity
h = height
p = pressure
ρ = fluid density

Though some physicists argue that it leads to the compromising of certain assumptions (i.e., incompressibility, no flow motivation, and a closed fluid loop), most agree it is correct to explain "lift" using Bernoulli's principle. This is because Bernoulli's principle can also be thought of as predicting that the pressure in moving fluid is less than the pressure in fluid at rest. Thus, there are many examples of physical phenomenon that can be explained by Bernoulli's Principle:

- The lift on airplane wings occurs because the top surface is curved while the bottom surface is straight. Air must therefore move at a higher velocity on the top of the wing and the resulting lower pressure on top accounts for lift.
- The tendency of windows to explode rather than implode in hurricanes is caused by the pressure drop that results from the high speed winds blowing across the outer surface of the window. The higher pressure on the inside of the window then pushes the glass outward, causing an explosion.
- The ballooning and fluttering of a tarp on the top of a semi-truck moving down the highway is caused by the flow of air across the top of the truck. The decrease in pressure causes the tarp to "puff up."
- A perfume atomizer pushes a stream of air across a pool of liquid. The drop in pressure caused by the moving air lifts a bit of the perfume and allows it to be dispensed.

DOMAIN II. ELECTRICITY AND MAGNETISM

Skill 1 Electric forces and Coulomb's law

Any point charge may experience force resulting from attraction to or repulsion from another charged object. The easiest way to begin analyzing this phenomenon and calculating this force is by considering two point charges. Let us say that the charge on the first point is Q_1, the charge on the second point is Q_2, and the distance between them is r. Their interaction is governed by **Coulomb's Law** which gives the formula for the force F as:

$$F = k \frac{Q_1 Q_2}{r^2}$$

where $k = 9.0 \times 10^9 \, \frac{N \cdot m^2}{C^2}$ (known as Coulomb's constant)

The charge is a scalar quantity, however, the force has direction. For two point charges, the direction of the force is along a line joining the two charges. Note that the force will be repulsive if the two charges are both positive or both negative and attractive if one charge is positive and the other negative. Thus, a negative force indicates an attractive force.

When more than one point charge is exerting force on a point charge, we simply apply Coulomb's Law multiple times and then combine the forces as we would in any statics problem. Let's examine the process in the following example problem.

<u>Problem:</u> Three point charges are located at the vertices of a right triangle as shown below. Charges, angles, and distances are provided (drawing not to scale). Find the force exerted on the point charge A.

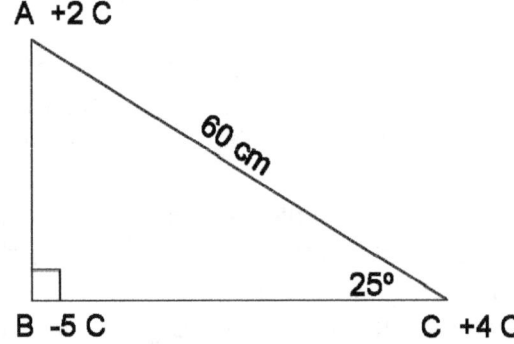

PHYSICS

Solution: First we find the individual forces exerted on A by point B and point C. We have the information we need to find the magnitude of the force exerted on A by C.

$$F_{AC} = k\frac{Q_1 Q_2}{r^2} = 9\times 10^9 \frac{N\cdot m^2}{C^2}\left(\frac{4C \times 2C}{(0.6m)^2}\right) = 2\times 10^{11} N$$

To determine the magnitude of the force exerted on A by B, we must first determine the distance between them.

$$\sin 25° = \frac{r_{AB}}{60cm}$$
$$r_{AB} = 60cm \times \sin 25° = 25cm$$

Now we can determine the force.

$$F_{AB} = k\frac{Q_1 Q_2}{r^2} = 9\times 10^9 \frac{N\cdot m^2}{C^2}\left(\frac{-5C \times 2C}{(0.25m)^2}\right) = -1.4\times 10^{12} N$$

We can see that there is an attraction in the direction of B (negative force) and repulsion in the direction of C (positive force). To find the net force, we must consider the direction of these forces (along the line connecting any two point charges). We add them together using the law of cosines.

$$F_A^2 = F_{AB}^2 + F_{AC}^2 - 2F_{AB}F_{AC}\cos 75°$$
$$F_A^2 = (-1.4\times 10^{12} N)^2 + (2\times 10^{11} N)^2 - 2(-1.4\times 10^{12} N)(2\times 10^{11} N)^2 \cos 75°$$
$$F_A = 1.5\times 10^{12} N$$

This gives us the magnitude of the net force, now we will find its direction using the law of sines.

$$\frac{\sin\theta}{F_{AC}} = \frac{\sin 75°}{F_A}$$
$$\sin\theta = F_{AC}\frac{\sin 75°}{F_A} = 2\times 10^{11} N \frac{\sin 75°}{1.5\times 10^{12} N}$$
$$\theta = 7.3°$$

Thus, the net force on A is 7.3° west of south and has magnitude 1.5×10^{12}N. Looking back at our diagram, this makes sense, because A should be attracted to B (pulled straight south) but the repulsion away from C "pushes" this force in a westward direction.

Skill 2   **Electric fields, Gauss's law, electric potential energy, electric potential, and potential difference**

An **electric field** exists in the space surrounding a charge. Electric fields have both direction and magnitude determined by the strength and direction in which they exhibit force on a test charge. The units used to measure electric fields are newtons per coulomb (N/C). **Electric potential** is simply the **potential energy** per unit of charge. Given this definition, it is clear that electric potential must be measured in joules per coulomb and this unit is known as a volt (J/C=V).

Within an electric field there are typically differences in potential energy. This **potential difference** may be referred to as **voltage**. The difference in electrical potential between two points is the amount of work needed to move a unit charge from the first point to the second point. Stated mathematically, this is:

$$V = \frac{W}{Q}$$

where V= the potential difference
W= the work done to move the charge
Q= the charge

We know from mechanics, however, that work is simply force applied over a certain distance. We can combine this with Coulomb's law to find the work done between two charges distance r apart.

$$W = F \cdot r = k\frac{Q_1 Q_2}{r^2} \cdot r = k\frac{Q_1 Q_2}{r}$$

Now we can simply substitute this back into the equation above for electric potential:

$$V_2 = \frac{W}{Q_2} = \frac{k\dfrac{Q_1 Q_2}{r}}{Q_2} = k\frac{Q_1}{r}$$

Let's examine a sample problem involving electrical potential.

Problem: What is the electric potential at point A due to the 2 shown charges? If a charge of +2.0 C were infinitely far away, how much work would be required to bring it to point A?

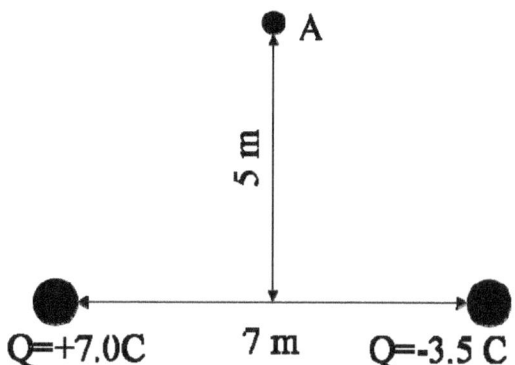

Solution: To determine the electric potential at point A, we simple find and add the potential from the two charges (this is the principle of superposition). From the diagram, we can assume that A is equidistant from each charge. Using the Pythagorean theorem, we determine this distance to be 6.1 m.

$$V = \frac{kq}{r} = k\left(\frac{7.0C}{6.1m} + \frac{-3.5C}{6.1m}\right) = 9 \times 10^9 \frac{N.m^2}{C^2}\left(0.57\frac{C}{m}\right) = 5.13 \times 10^9 V$$

Now, let's consider bringing the charged particle to point A. We assume that electric potential of these particle is initially zero because it is infinitely far away. Since now know the potential at point A, we can calculate the work necessary to bring the particle from V=0, i.e. the potential energy of the charge in the electrical field:

$$W = VQ = (5.13 \times 10^9) \times 2J = 10.26 \times 10^9 J$$

The large results for potential and work make it apparent how large the unit coulomb is. For this reason, most problems deal in microcoulombs (μC).

In any physical phenomenon, flux refers to rate of movement of a substance or energy through a certain area. Flux can be used to quantify the movement of mass, heat, momentum, light, molecules and other things. Flux depends on density of flow, area, and direction of the flow. To visualize this, imagine a kitchen sieve under a tap of flowing water. The water that passes through the sieve is the flux; the flux will decrease if we lower the water flow rate, decrease the size of the sieve, or tilt the sieve away from direction of the water's flow. Electric flux, then, is just the number of electric field lines that pass through a given area. It is given by the following equation:

$$\Phi = E(\cos\phi)A$$

where Φ = flux
E = the electric field
A = area
ϕ = the angle between the electric field and a vector normal to the surface A

Thus, if a plane is parallel to an electric field, no field lines will pass through that plane and the flux through it will be zero. If a plane is perpendicular to an electric field, the flux through it will be maximal.

Gauss's Law says that the electric flux through a surface is equal to the charge enclosed by a surface divided by a constant ε_0 (permittivity of free space). The simplest mathematical statement of this law is:

$$\Phi = Q_A/\varepsilon_0$$

where Q_A = the charge enclosed by the surface

Gauss's Law provides us with a useful and powerful method to calculate electric fields. For instance, imagine a solid conducting sphere with a net charge Q_s. We know from Gauss's Law that the electric field inside the sphere must be zero and all the excess charge lies on the outer surface of the sphere. The field produced by this sphere is the same a point charge of value Q_s. This conclusion is true whether the sphere is solid or hollow.

Skill 3 Conductors, insulators and semiconductors (charging by friction, conduction and induction)

Electrical current requires the free flow of electrons. Various materials allow different degrees of electron movement and are classified as conductors, insulators, or semiconductors (in certain, typically man-made environments, superconductors also exist). When charge is transferred to a mass of material, the response is highly dependent on whether that material is a conductor or insulator.

Conductors: Those materials that allow for free and easy movement of electrons are called conductors. Some of the best conductors are metal, especially copper and silver. This is because these materials are held together with metallic bonds, which involve de-localized electrons shared by atoms in a lattice. If a charge is transferred to a conductor, the electrons will flow freely and the charge will quickly distribute itself across the material in a manner dictated by the conductor's shape.

Insulators: Materials that do not allow conduction are call insulators. Good insulators include, glass, rubber, and wood. These materials have chemical structures in which the electrons are closely localized to the individual atoms. In contrast to a conductor, a charge transferred to an insulator will remain localized at the point where it was introduced. This is because the movement of electrons will be highly impeded.

Semiconductors: Materials with intermediate conduction properties are known as semiconductors. Their properties are similar to insulators in that they have few free electrons to carry the charge. However, these electrons can be thermally excited into higher energy states that allow them sufficient freedom to transmit electrical charge. The electrical properties of a semiconductor are often improved by introducing impurities, a procedure known as doping. Doping introduces extra electrons or extra electron acceptors to facilitate the movement of charge. *Temporary* changes in electrical properties can be induced by applying an electrical field.

Charge can be transferred to materials in a few different ways:

Conduction: In the most general sense, electrical conduction is the movement of charged particles (electrons) through a medium. As explained above, in conducting materials, electrons loosely attached to atoms are capable of carrying an electrical current. This requires that the atoms of the conducting material be brought into physical contact with the charge source. For example, if a conductor is brought in contact with another charged conductor or a current source (such as a battery), the charge will be transferred to that conductor.

Friction: When two materials are brought into contact (or rubbed against each other), electrons may be transferred between them. If the materials are then separated, one will be left with a negative charge and the other will have a positive charge. This phenomenon is known as the triboelectric effect. Both the polarity and strength of the resultant charge depends upon the materials, surface roughness, temperature, and strain. Some common examples of combinations that produce significant charges are glass with silk and rubber with fur.

Induction: Electromagnetic induction is the production of voltage that occurs when a conductor interacts with a magnetic field. A conductor can be charged either by moving it through a static electric field or by placing it in a changing magnetic field. Induction was discovered by Michael Faraday and Faraday's Law governs this phenomenon (see section II.16).

Skill 4 Capacitance and dielectrics

Capacitance (C) is a measure of the stored electric charge per unit electric potential. The mathematical definition is:

$$C = \frac{Q}{V}$$

It follows from the definition above that the units of capacitance are coulombs per volt, a unit known as a farad ($F=C/V$). In circuits, devices called parallel plate capacitors are formed by two closely spaced conductors. The function of capacitors is to store electrical energy. When a voltage is applied, electrical charges build up in both the conductors (typically referred to as plates). These charges on the two plates have equal magnitude but opposite sign. The capacitance of a capacitor is a function of the distance d between the two plates and the area A of the plates:

$$C \approx \frac{\varepsilon A}{d}; A \gg d^2$$

Capacitance also depends on the permittivity of the non-conducting matter between the plates of the capacitor. This matter may be only air or almost any other non-conducting material and is referred to as a **dielectric**. The permittivity of empty space ε_0 is roughly equivalent to that for air, $\varepsilon_{air}=8.854 \times 10^{-12}$ C^2/N•m^2. For other materials, the dielectric constant, κ, is the permittivity of the material in relation to air ($\kappa=\varepsilon/\varepsilon_{air}$). The make-up of the dielectric is critical to the capacitor's function because it determines the maximum energy that can be stored by the capacitor. This is because an overly strong electric field will eventually destroy the dielectric.

In summary, a capacitor is "charged" as electrical energy is delivered to it and opposite charges accumulate on the two plates. The two plates generate electric fields and a voltage develops across the dielectric. The energy stored in the capacitor, then, is equal to the amount of work necessary to create this voltage. The mathematical statement of this is:

$$E_{stored} = \frac{1}{2}CV^2 = \frac{1}{2}\frac{Q^2}{C} = \frac{1}{2}VQ$$

The work per unit volume or the **electric field energy density** within a capacitor can be shown to be $\eta = \frac{1}{2}\varepsilon E^2$. This result is generally valid for the energy per unit volume of any electrostatic field, not only for a constant field within a capacitor.

Problem: Imagine that a parallel plate capacitor has an area of 10.00 cm^2 and a capacitance of 4.50 pF. The capacitor is connected to a 12.0 V battery. The capacitor is completely charged and then the battery is removed. What is the separation of the plates in the capacitor? How much energy is stored between the plates? We've assumed that this capacitor initially had no dielectric (i.e., only air between the plates) but now imagine it has a Mylar dielectric that fully fills the space. What will the new capacitance be? (for Mylar, ◻=3.5)

Solution: To determine the separation of the plates, we use our equation for a capacitor:

$$C = \frac{\varepsilon_0 A}{d}$$

We can simply solve for d and plug in our values:

$$d = \varepsilon_0 \frac{A}{C} = \left(8.854 \times 10^{-12} \frac{C}{N \cdot m^2}\right) \frac{10 \times 10^{-4} m^2}{4.5 \times 10^{-12} F} = 1.97 \times 10^{-3} m = 1.97 mm$$

Similarly, to find stored energy, we simply employ the equation above:

$$E_{stored} = \frac{1}{2}QV$$

But we don't yet know the charge Q, so we must first find it from the definition of capacitance:

$$C = \frac{Q}{V}$$
$$Q = CV = (4.5 \times 10^{-12}) \times (12V) = 5.4 \times 10^{-11} C$$

Now we can find the stored energy:

$$E_{stored} = \frac{1}{2}QV = \frac{1}{2}(5.4 \times 10^{-11} C)(12V) = 3.24 \times 10^{-10} J$$

To find the capacitance with a Mylar dielectric, we again use the equation for capacitance of a parallel plate capacitor. Note that the new capacitance can be found by multiplying the original capacitance by κ:

$$C = \frac{\kappa_{Mylar}\varepsilon_0 A}{d} = \kappa_{Mylar}C_0 = 3.5 \times 4.5\,pF = 15.75\,pF$$

Skill 5 Conductors, insulators, and semiconductors as used in circuits

Conductors: Conductors are extremely important in circuits since they are used whenever a charge must be carried from one place to another. Wires, resistors, and inductors are made from conducting materials. On a larger scale, electrical lines that carry power are made from conductors. Finally, the plates of a capacitor and the contacts of a switch must be made from conducting materials. In circuit analysis, we use Ohm's Law (V=IR) to study how current flows through a conducting material.

Insulators: Insulators are especially useful for containing or stopping the flow of current. So they are used in handles and switches and any other places where stopping the flow of current is necessary. Additionally, the dielectrics within capacitors are made of insulators.

Semiconductors: Semi-conductors are useful in situations where controlled conduction is required. Often made from primarily silicon, semiconductors are key components in the circuitry in modern electronics (cell phones, computers, and other digital devices).

Skill 6 Sources of EMF (batteries, photocells, generators)

A **battery** produces **electrical energy from chemical energy**. Typical batteries have two terminals, a positive and a negative one. Electrochemical reactions take place in the **electrolyte**, a chemical stored within a battery, producing electrons that accumulate on the negative terminal or the **cathode**. If the positive and negative terminals are connected, these electrons flow to the **anode** (positive terminal) creating an electrical current in the circuit. The first battery created by Alessandro Volta in 1800 was known as the voltaic pile. Many different types of batteries are available today for a variety of applications. Battery electrolytes may be **acidic** or **alkaline**. Automobile batteries, for example, are acidic. Alkaline batteries are used in devices such as cell phones, radios and pagers. Batteries may also be classified as **wet** or **dry** depending on the nature of the electrolyte. The chemical energy of a **rechargeable** battery can be replenished and the battery reused.

A **photocell** uses the **photoelectric effect** (see section V.8), emission of electrons by atoms that absorb light photons, to generate electricity. A type of photocell, the **photovoltaic cell** (e.g. solar cell) converts **light energy into electrical energy**. A simple photovoltaic cell consists of light-sensitive semiconducting material placed between two electrodes that are connected in a circuit. Electrons released by a light incident on the cell create a current in the circuit and can be harnessed to operate electrical devices.

Generators are devices that are convert **mechanical energy into electrical energy**. The mechanical energy can come from a variety of sources; combustion engines, blowing wind, falling water, or even a hand crank or bicycle wheel. Most generators rely upon electromagnetic induction to create an electrical current. These generators basically consist of magnets and a coil. The magnets create a magnetic field and the coil is located within this field. Mechanical energy, from whatever source, is used to spin the coil within this field. As stated by Faraday's Law, this produces a voltage. It is important to understand that generators move electric current, but do not create electrical charge. We can make an analogy to a water pump, which can create water current, but does not create water.

Skill 7   Current and resistance (Ohm's law, resistivity)

Ohm's Law is the most important tool we posses to analyze electrical circuits. Ohm's Law states that the current passing through a conductor is directly proportional to the voltage drop and inversely proportional to the resistance of the conductor. Stated mathematically, this is:

$$V=IR$$

Problem:
The circuit diagram at right shows three resistors connected to a battery in series. A current of 1.0A flows through the circuit in the direction shown. It is known that the equivalent resistance of this circuit is 25 Ω. What is the total voltage supplied by the battery?

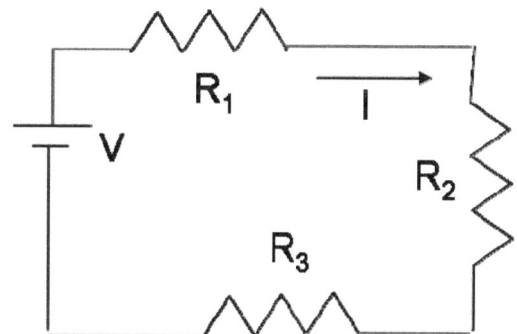

Solution:

To determine the battery's voltage, we simply apply Ohm's Law:

$$V = IR = 1.0A \times 25\Omega = 25V$$

Conductors are those materials which allow for the free passage of electrical current. However, all materials exhibit a certain opposition to the movement of electrons. This opposition is known as **resistivity** (ρ). Resistivity is determined experimentally by measuring the resistance of a uniformly shaped sample of the material and applying the following equation:

$$\rho = R\frac{A}{l}$$

where ρ = static resistivity of the material
R = electrical resistance of the material sample
A = cross-sectional area of the material sample
L = length of the material sample

The temperature at which these measurements are taken is important as it has been shown that resistivity is a function of temperature. For conductors, resistivity increases with increasing temperature and decreases with decreasing temperature. At extremely low temperatures resistivity assumes a low and constant value known as residual resistivity (ρ_0). Residual resistivity is a function of the type and purity of the conductor.

The following equation allows us to calculate the resistivity ρ of a material at any temperature given the resistivity at a reference temperature, in this case at $20^\circ C$:

$$\rho = \rho_{20}[1 + a(t - 20)]$$

where ρ_{20} = resistivity at $20^\circ C$

a = proportionality constant characteristic of the material

t = temperature in Celsius

Problem: The tungsten filament in a certain light bulb is a wire 8 μm in diameter and 10 mm long. Given that, for tungsten, $\rho_{20} = 5.5 \times 10^{-8}$ Ω·m and $a = 4.5 \times 10^{-3} K^{-1}$, what will the resistance of the filament be at 45°C?

Solution: First we must find the resistivity of the tungsten at 45°C:

$$\rho = 5.5 \times 10^{-8}(1 + 4.5 \times 10^{-3}(45 - 20)) = 6.1 \times 10^{-8} \Omega \cdot m$$

Now we can rearrange the equation defining resistivity and solve for the resistance of the filament:

$$R = \rho\frac{l}{A} = 6.1 \times 10^{-8} \times 0.01 / (\pi(4 \times 10^{-6})^2) = 12.1 \Omega$$

Skill 8   Capacitance and inductance

Section II.4 provides an extensive discussion of capacitance. Inductance L measures the amount of magnetic flux Φ created by a current i and is given by $L = \dfrac{\Phi}{i}$. The unit of inductance is Henrys when the flux is expressed in Webers and the current in Amperes. While a capacitor resists a change in voltage through a circuit, an inductor resists a change in current. As indicated by Faraday's law, the magnetic flux generated by the inductor creates an emf that opposes the flow of current.

In a DC current circuit, a capacitor behaves like an open circuit (infinite resistance) while an inductor behaves like a short circuit (zero resistance) in the steady state following the initial transient response (where a capacitor builds up charge and an inductor builds up a magnetic field).

In an AC circuit, however, both capacitors and inductors contribute to the net **impedance** in the circuit which is a measure of opposition to the alternating current. It is similar to resistance and also has the unit ohm. However, due to the phased nature of AC, impedance is a complex number, having both real and imaginary components. The resistance is the real part of impedance while the **reactance** of capacitors and inductors constitute the imaginary part. The relationship between impedance (Z), resistance (R), and reactance (X) is given by below.

$$Z = R + Xi \quad \text{(Remember that } i=\sqrt{-1}\text{)}$$

The inductive reactance is given by $X_L = \omega L$ where ω is the angular frequency of the current and L is the inductance. The capacitive reactance is given by $X_C = \dfrac{1}{\omega C}$ where ω is the angular frequency of the current and C is the capacitance.

Skill 9   Energy and power

Electrical power is a measure of energy used or available per unit time, i.e. how much work can be done by an electrical current in unit time. For descriptions of electrical potential energy, energy stored in a capacitor and electric field energy density see sections II.2 and II.4.

Electrical power has units of watts (W), i.e. Joules/second. To determine electrical power, we simply use **Joule's law**:

$$P = IV$$

where P=power
I=current
V=voltage

If we combine this with **Ohm's law** (V=IR), we generate two new equations that are useful for finding the amount of power dissipated by a resistor:

$$P = I^2 R$$

$$P = \frac{V^2}{R}$$

Problem: How much power is dissipated by a 1 kΩ resistor with a 50V voltage drop through it?

Solution: $P = \dfrac{V^2}{R} = \dfrac{(50V)^2}{1000\Omega} = 2.5W$

Skill 10   **Analyzing circuits (series and parallel circuits using Ohm's law or Kirchhoff's rules, resistors and capacitors in series or parallel, internal resistance, RC circuits)**

Kirchoff's Laws are a pair of laws that apply to conservation of charge and energy in circuits and were developed by Gustav Kirchoff.

Kirchoff's Current Law: At any point in a circuit where charge density is constant, the sum of currents flowing toward the point must be equal to the sum of currents flowing away from that point.

Kirchoff's Voltage Law: The sum of the electrical potential differences around a circuit must be zero.

While these statements may seem rather simple, they can be very useful in analyzing DC circuits, those involving constant circuit voltages and currents.

Problem:
The circuit diagram at right shows three resistors connected to a battery in series. A current of 1.0 A is generated by the battery. The potential drop across R_1, R_2, and R_3 are 5V, 6V, and 10V. What is the total voltage supplied by the battery?

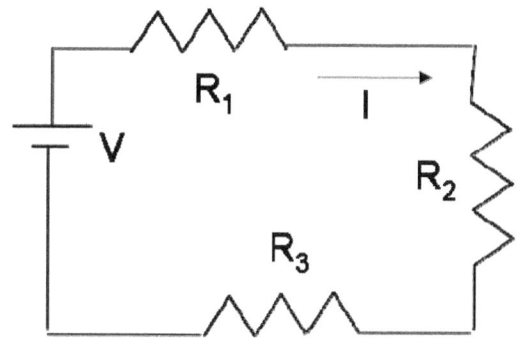

Solution:
Kirchoff's Voltage Law tells us that the total voltage supplied by the battery must be equal to the total voltage drop across the circuit. Therefore:

$$V_{battery} = V_{R_1} + V_{R_2} + V_{R_3} = 5V + 6V + 10V = 21V$$

Problem:
The circuit diagram at right shows three resistors wired in parallel with a 12V battery. The resistances of R_1, R_2, and R_3 are 4 Ω, 5 Ω, and 6 Ω, respectively. What is the total current?

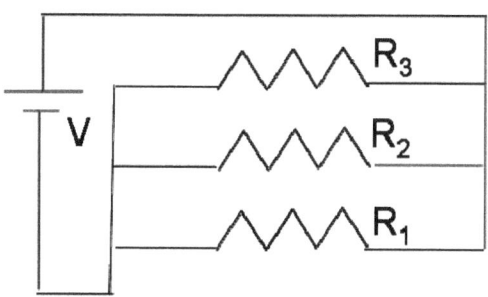

Solution:
This is a more complicated problem. Because the resistors are wired in parallel, we know that the voltage entering each resistor must be the same and equal to that supplied by the battery. We can combine this knowledge with **Ohm's Law** to determine the current across each resistor:

$$I_1 = \frac{V_1}{R_1} = \frac{12V}{4\Omega} = 3A$$

$$I_2 = \frac{V_2}{R_2} = \frac{12V}{5\Omega} = 2.4A$$

$$I_3 = \frac{V_3}{R_3} = \frac{12V}{6\Omega} = 2A$$

Finally, we use Kirchoff's Current Law to find the total current:

$$I = I_1 + I_2 + I_3 = 3A + 2.4A + 2A = 7.4A$$

Resistors and capacitors are often used together in series or parallel. Two components are in series if one end of the first element is connected to one end of the second component. The components are in parallel if both ends of one element are connected to the corresponding ends of another. A series circuit has a single path for current flow through all of its elements. A parallel circuit is one that requires more than one path for current flow in order to reach all of the circuit elements.

Below is a diagram demonstrating a simple circuit with resistors in parallel (on right) and in series (on left). Note the symbols used for a battery (noted V) and the resistors (noted R).

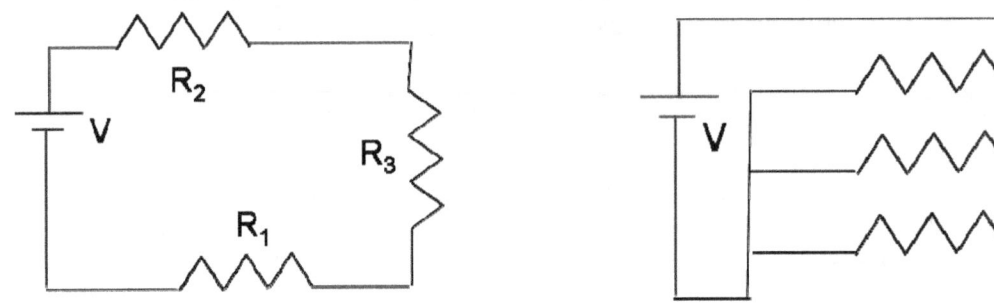

Thus, when the resistors are placed in series, the current through each one will be the same. When they are placed in parallel, the voltage through each one will be the same. To understand basic circuitry, it is important to master the rules by which the equivalent resistance (R_{eq}) or capacitance (C_{eq}) can be calculated from a number of resistors or capacitors:

Resistors in parallel: $$\frac{1}{R_{eq}} = \frac{1}{R_1} + \frac{1}{R_2} + \cdots + \frac{1}{R_n}$$

Resistors in series: $$R_{eq} = R_1 + R_2 + \cdots + R_n$$

Capacitors in parallel: $$C_{eq} = C_1 + C_2 + \cdots + C_n$$

Capacitors in series: $$\frac{1}{C_{eq}} = \frac{1}{C_1} + \frac{1}{C_2} + \cdots + \frac{1}{C_n}$$

In the example problems above, we have assumed that the voltage across the terminals of a battery is equal to its emf. In reality, the terminal voltage decreases slightly as the current increases. This decrease is due to the **internal resistance** r of the battery. One way to think of the internal resistance is as a small resistor in series with an ideal battery. Thus, terminal voltage drop is given by $V = \varepsilon - Ir$ where ε is the emf of the battery, r is the internal resistance and I is the current flowing through the circuit.

Resistors and capacitors may also be combined together in circuits. Below we will consider the simplest RC circuit with a resistor and a capacitor in series connected to a voltage source.

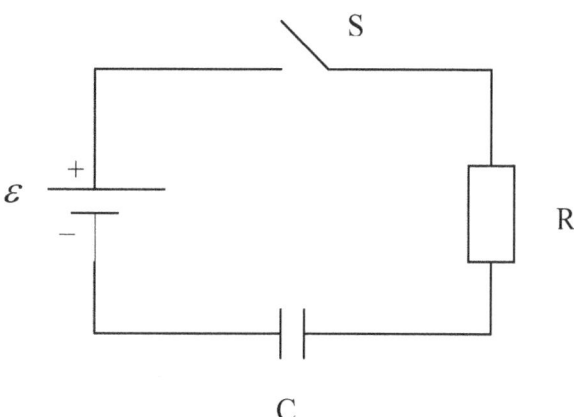

Applying Kirchhoff's first rule we get $\varepsilon = V_R + V_C = IR + \dfrac{Q}{C} = R\dfrac{dQ}{dt} + \dfrac{Q}{C}$ since the current I in the circuit is equal to the rate of increase of charge on the capacitor. If the charge on the capacitor is zero when the switch is closed at time t=0, the current I in the circuit starts out at the value ε/R and gradually falls to zero as the capacitor charge gradually rises to its maximum value. The mathematical expressions for these two quantities are

$$Q(t) = C\varepsilon(1 - e^{-t/RC}); I(t) = \dfrac{\varepsilon}{R}e^{-t/RC}$$

The product RC is known as the **time constant** of the circuit.

If the battery is now removed and the switch closed at t=0, the capacitor will slowly discharge with its charge at any point t given by

$Q(t) = Q_f e^{-t/RC}$ where Q_f is the charge that the capacitor started with at t=0.

Skill 11   Power in alternating-current circuits (average power and energy transmission)

Alternating current (AC) is a type of electrical current with cyclically varying magnitude and direction. This is differentiated from direct current (DC), which has constant direction. AC is the type of current delivered to businesses and residences.

Though other waveforms are sometimes used, the vast majority of AC current is sinusoidal. Thus we can use wave terminology to help us describe AC current. Since AC current is a function of time, we can express it mathematically as:

$$v(t) = V_{peak} \cdot \sin(\omega t)$$

where V_{peak}= the peak voltage; the maximum value of the voltage

ω=angular frequency; a measure of rotation rate

t=time

The instantaneous power or energy transmission per unit time is given by $I^2_{peak} R \sin^2(\omega t)$, a value that varies over time. In order to asses the overall rate of energy transmission, however, we need some kind of average value. The **root mean square value (V_{rms}, I_{rms})** is a specific type of average given by the following formulae:

$$V_{rms} = \frac{V_{peak}}{\sqrt{2}} \quad ; \quad I_{rms} = \frac{I_{peak}}{\sqrt{2}} ; \quad I_{rms} = \frac{V_{rms}}{R}$$

V_{rms} is useful because an AC current will deliver the same power as a DC current if its V_{rms}=V_{DC}, i.e. average power or average energy transmission per unit time is given by $P_{av} = V_{rms} I_{rms}$.

Skill 12 **Measurement of potential difference, current, resistance, and capacitance (ammeter, galvanometer, voltmeter and potentiometer)**

Electrical meters function by utilizing the following familiar equations:

Across a resistor (Resistor R):

$$V_R = IR_R$$

Across a capacitor (Capacitor C):

$$V_C = IX_C$$

Across an inductor (Inductor L):

$$V_L = IX_L$$

Where V=voltage, I=current, R=resistance, X=reactance.

Ammeter: An ammeter placed in series in a circuit measures the current through the circuit. An ammeter typically has a very small resistance so that the current in the circuit is not changed too much by insertion of the ammeter.

Voltmeter: A voltmeter is used to measure potential difference. The potential difference across a resistor is measured by a voltmeter placed in parallel across it. An ideal voltmeter has very high resistance so that it does not appreciably alter circuit resistance and therefore the voltage drop it is measuring.

Galvanometer: A galvanometer is a device that measures current and is a component of an ammeter or a voltmeter. A typical galvanometer consists of a coil of wire, an indicator and a scale that is designed to be proportional to the current in the galvanometer. The principle that a current-carrying wire experiences a force in a magnetic field is used in the construction of a galvanometer. In order to create a voltmeter from a galvanometer, resistors are added in series to it. To build an ammeter using a galvanometer, a small resistance known as a **shunt resistor** is placed in parallel with it.

Potentiometer: A potentiometer is a variable resistance device in which the user can vary the resistance to control the current and voltage applied to a circuit. Since the potentiometer can be used to control what fraction of the emf of a battery is applied to a circuit, it is also known as a voltage divider. It can be used to measure an unknown voltage by comparing it with a known value.

Multimeter: A common electrical meter, typically known as a multimeter, is capable of measuring voltage, resistance, and current. Many of these devices can also measure capacitance (farads), frequency (hertz), duty cycle (a percentage), temperature (degrees), conductance (siemens), and inductance (henrys).

Skill 13   Magnets, magnetic fields, and magnetic forces (magnetic dipoles and materials, forces on a charged particle moving in a magnetic and/or electric field (Lorentz force law, cyclotron, mass spectrometer) forces or torques on current carrying conductors in magnetic fields)

Magnetism is a phenomenon in which certain materials, known as magnetic materials, attract or repel each other. A magnet has two poles, a south pole and a north pole. Like poles repel while unlike poles attract. Magnetic poles always occur in pairs known as **magnetic dipoles**. One cannot isolate a single magnetic pole. If a magnet is broken in half, opposite poles appear at both sides of the break point so that one now has two magnets each with a south pole and a north pole. No matter how small the pieces a magnet is broken into, the smallest unit at the atomic level is still a dipole.

A large magnet can be thought of as one with many small dipoles that are aligned in such a way that apart from the pole areas, the internal south and north poles cancel each other out. Destroying this long range order within a magnet by heating or hammering can demagnetize it. The dipoles in a **non-magnetic** material are randomly aligned while they are perfectly aligned in a preferred direction in **permanent** magnets. In a **ferromagnet**, there are domains where the magnetic dipoles are aligned, however, the domains themselves are randomly oriented. A ferromagnet can be magnetized by placing it in an external magnetic field that exerts a force to line up the domains.

A magnet produces a magnetic field that exerts a force on any other magnet or current-carrying conductor placed in the field. Magnetic field lines are a good way to visualize a magnetic field. The distance between magnetic fields lines indicates the strength of the magnetic field such that the lines are closer together near the poles of the magnets where the magnetic field is the strongest. The lines spread out above and below the middle of the magnet, as the field is weakest at those points furthest from the two poles. The SI unit for magnetic field known as magnetic induction is Tesla(T) given by $1T = 1 \text{ N.s}/(\text{C.m}) = 1 \text{ N}/(\text{A.m})$. Magnetic fields are often expressed in the smaller unit Gauss (G) ($1 \text{ T} = 10{,}000$ G). Magnetic field lines always point from the north pole of a magnet to the south pole.

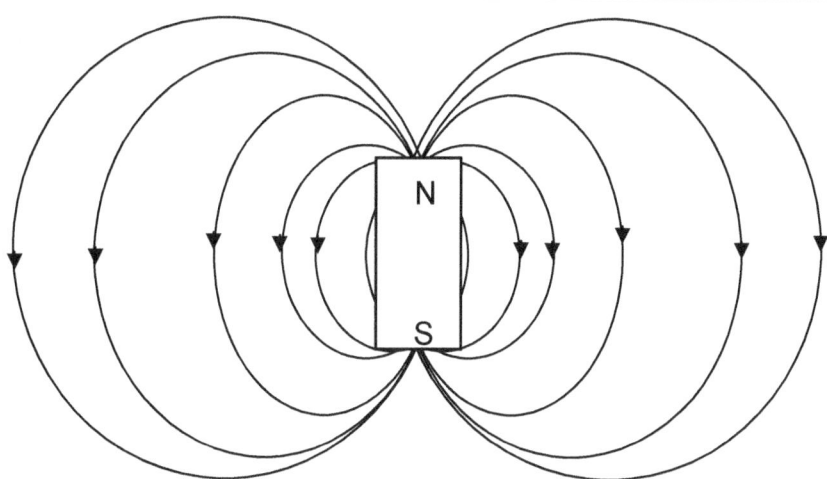

Magnetic field lines can be plotted with a magnetized needle that is free to turn in 3 dimensions. Usually a compass needle is used in demonstrations. The direction tangent to the magnetic field line is the direction the compass needle will point in a magnetic field. Iron filings spread on a flat surface or magnetic field viewing film which contains a slurry of iron filings are another way to see magnetic field lines.

The magnetic force exerted on a charge moving in a magnetic field depends on the size and velocity of the charge as well as the magnitude of the magnetic field. One important fact to remember is that only the velocity of the charge in a direction perpendicular to the magnetic field will affect the force exerted. Therefore, a charge moving parallel to the magnetic field will have no force acting upon it whereas a charge will feel the greatest force when moving perpendicular to the magnetic field.

The direction of the magnetic force, or the magnetic component of the **Lorenz force** (force on a charged particle in an electrical and magnetic field), is always at a right angle to the plane formed by the velocity vector v and the magnetic field B and is given by applying the right hand rule - if the fingers of the right hand are curled in a way that seems to rotate the v vector into the B vector, the thumb points in the direction of the force. The magnitude of the force is equal to the cross product of the velocity of the charge with the magnetic field multiplied by the magnitude of the charge.

$$F = q\,(\mathbf{v} \times \mathbf{B}) \quad or \quad F = q\,v\,B\sin(\theta)$$

Where θ is the angle formed between the vectors of velocity of the charge and direction of magnetic field.

Problem: Assuming we have a particle of 1×10^{-6} kg that has a charge of -8 coulombs that is moving perpendicular to a magnetic field in a clockwise direction on a circular path with a radius of 2 m and a speed of 2000 m/s, let's determine the magnitude and direction of the magnetic field acting upon it.

Solution: We know the mass, charge, speed, and path radius of the charged particle. Combining the equation above with the equation for centripetal force we get

$$qvB = \frac{mv^2}{r} \quad \text{or} \quad B = \frac{mv}{qr}$$

Thus B= (1 x 10⁻⁶ kg) (2000m/s) / (-8 C)(2 m) = 1.25 x 10⁻⁴ Tesla

Since the particle is moving in a clockwise direction, we use the right hand rule and point our fingers clockwise along a circular path in the plane of the paper while pointing the thumb towards the center in the direction of the centripetal force. This requires the fingers to curl in a way that indicates that the magnetic field is pointing out of the page. However, since the particle has a negative charge we must reverse the final direction of the magnetic field into the page.

A **mass spectrometer** measures the mass to charge ratio of ions using a setup similar to the one described above. m/q is determined by measuring the path radius of particles of known velocity moving in a known magnetic field.

A **cyclotron**, a type of particle accelerator, also uses a perpendicular magnetic field to keep particles on a circular path. After each half circle, the particles are accelerated by an electric field and the path radius is increased. Thus the beam of particles moves faster and faster in a growing spiral within the confines of the cyclotron until they exit at a high speed near the outer edge. Its compactness is one of the advantages a cyclotron has over linear accelerators.

The **force on a current-carrying conductor** in a magnetic field is the sum of the forces on the moving charged particles that create the current. For a current I flowing in a straight wire segment of length l in a magnetic field B, this force is given by

$$\mathbf{F} = I\mathbf{l} \times \mathbf{B}$$

where *l* is a vector of magnitude l and direction in the direction of the current.

When a current-carrying loop is placed in a magnetic field, the net force on it is zero since the forces on the different parts of the loop act in different directions and cancel each other out. There is, however, a net torque on the loop that tends to rotate it so that the area of the loop is perpendicular to the magnetic field. For a current I flowing in a loop of area A, this torque is given by

$$\tau = IA\hat{\mathbf{n}} \times \mathbf{B}$$

where $\hat{\mathbf{n}}$ is the unit vector perpendicular to the plane of the loop.

Magnetic flux (Gauss's law of magnetism)

Carl Friedrich Gauss developed laws that related electric or gravitational flux to electrical charge or mass, respectively. Gauss's law, along with others, was eventually generalized by James Clerk Maxwell to explain the relationships between electromagnetic phenomena (Maxwell's Equations).

To understand Gauss's law for magnetism, we must first define magnetic flux. Magnetic flux is the magnetic field that passes through a given area. It is given by the following equation:

$$\Phi = B(\cos\phi)A$$

where Φ=flux
B=the magnetic field
A=area
ϕ= the angle between the electric field and a vector normal to the surface A

Thus, if a plane is parallel to a magnetic field, no magnetic field lines will pass through that plane and the flux will be zero. If a plane is perpendicular to a magnetic field, the flux will be maximal.

Now we can state Gauss's law of magnetism: the net magnetic flux out of any closed surface is zero. Mathematically, this may be stated as:

$$\vec{\nabla} \cdot \vec{B} = 0$$

where $\vec{\nabla}$ = the del operator
$\vec{B}$ = magnetic field

One of the most important implications of this law is that there are no magnetic monopoles (that is, magnets always have a positive and negative pole). This is because a magnetic monopole source would give a non-zero product in the equation above. For a magnetic dipole with closed surface, of course, the net flux will always be zero. This is because the magnetic flux directed inward toward the south pole is always equal to the magnetic flux outward from the north pole.

Skill 14   **Magnetic fields produced by currents (Biot-Savart law, Ampere's law, magnetic field of a wire, magnetic field of a solenoid, displacement current)**

Conductors through which electrical currents travel will produce magnetic fields: The magnetic field dB induced at a distance r by an element of current Idl flowing through a wire element of length dl is given by the **Biot-Savart** law

$$dB = \frac{\mu_0}{4\pi} \frac{Idl \times \hat{r}}{r^2}$$

where μ_0 is a constant known as the permeability of free space and $\hat{r}$ is the unit vector pointing from the current element to the point where the magnetic field is calculated.

An alternate statement of this law is **Ampere's law** according to which the line integral of $B.dl$ around any closed path enclosing a steady current I is given by

$$\oint_C B \cdot dl = \mu_0 I$$

The basis of this phenomenon is the same no matter what the shape of the conductor, but we will consider three common situations:

Straight Wire
Around a current-carrying straight wire, the magnetic field lines form concentric circles around the wire. The direction of the magnetic field is given by the right-hand rule: When the thumb of the right hand points in the direction of the current, the fingers curl around the wire in the direction of the magnetic field. Note the direction of the current and magnetic field in the diagram.

To find the magnetic field of an infinitely long (allowing us to disregarding end effects) we apply Ampere's Law to a circular path at a distance r around the wire:

$$B = \frac{\mu_0 I}{2\pi r}$$

where μ_0=the permeability of free space ($4\pi \times 10^{-7}$ T·m/A)
I=current
r=distance from the wire

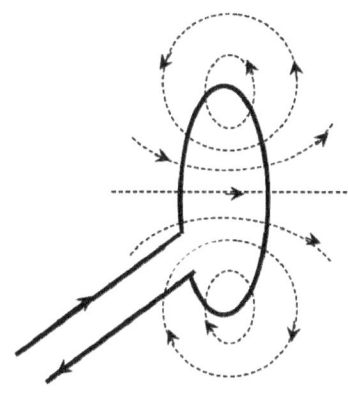

Loops
Like the straight wire from which it's been made, a looped wire has magnetic field lines that form concentric circles with direction following the right-hand rule. However, the field are additive in the center of the loop creating a field like the one shown. The magnetic field of a loop is found similarly to that for a straight wire.

In the center of the loop, the magnetic field is:

$$B = \frac{\mu_0 I}{2r}$$

Solenoids
A solenoid is essentially a coil of conduction wire wrapped around a central object. This means it is a series of loops and the magnetic field is similarly a sum of the fields that would form around several loops, as shown.

The magnetic field of a solenoid can be found as with the following equation:

$$B = \mu_0 n I$$

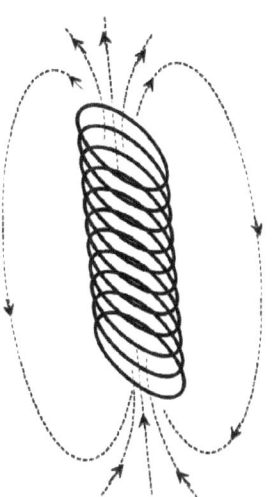

In this equation, n is turn density, which is simply the number of turns divided by the length of the solenoid.

Displacement current

While Ampere's law works perfectly for a steady current, for a situation where the current varies and a charge builds up (e.g. charging of a capacitor) it does not hold. Maxwell amended Ampere's law to include an additional term that includes the displacement current. This is not a true current but actually refers to changes in the electric field and is given by

$$I_d = \varepsilon_0 \frac{d\varphi_e}{dt}$$

where φ_e is the flux of the electric field. Including the displacement current, Ampere's law is given by

$$\oint \mathbf{B}.d\mathbf{l} = \mu_0 I + \mu_0 \varepsilon_0 \frac{d\varphi_e}{dt}$$

The displacement current essentially indicates that changing electric flux produces a magnetic field.

Skill 15   **Electromagnetic induction (magnetic flux, Lenz's law, Faraday's law, transformers, generators and motors)**

When the magnetic flux through a coil is changed, a voltage is produced which is known as induced electromagnetic force. Magnetic flux is a term used to describe the number of magnetic fields lines that pass through an area and is described by the equation:

$$\Phi = B A \cos\theta$$

Where Φ is the angle between the magnetic field B, and the normal to the plane of the coil of area A

By changing any of these three inputs, magnetic field, area of coil, or angle between field and coil, the flux will change and an EMF can be induced. The speed at which these changes occur also affects the magnitude of the EMF, as a more rapid transition generates more EMF than a gradual one. This is described by **Faraday's law** of induction:

$$\varepsilon = -N \Delta\Phi / \Delta t$$

where ε is emf induced, N is the number of loops in a coil, t is time, and Φ is magnetic flux

The negative sign signifies **Lenz's law** which states that induced emf in a coil acts to oppose any change in magnetic flux. Thus the current flows in a way that creates a magnetic field in the direction opposing the change in flux. The right hand rule for this is that if your fingers curl in the direction of the induced current, your thumb points in the direction of the magnetic field it produces through the loop.

Consider a coil lying flat on the page with a square cross section that is 10 cm by 5 cm. The coil consists of 10 loops and has a magnetic field of 0.5 T passing through it coming out of the page. Let's find the induced EMF when the magnetic field is changed to 0.8 T in 2 seconds.

First, let's find the initial magnetic flux: Φ_i

$\Phi_i = BA \cos\theta = (.5 \text{ T}) (.05 \text{ m}) (.1 \text{m}) \cos 0° = 0.0025 \text{ T m}^2$

And the final magnetic flux: Φ_f

$\Phi_f = BA \cos\theta = (0.8 \text{ T}) (.05 \text{ m}) (.1 \text{m}) \cos 0° = 0.004 \text{ T m}^2$

The induced emf is calculated then by

$\varepsilon = -N \Delta\Phi / \Delta t = -10 (.004 \text{ T m}^2 - .0025 \text{ T m}^2) / 2 \text{ s} = -0.0075$ volts.

To determine the direction the current flows in the coil we need to apply the right hand rule and Lenz's law. The magnetic flux is being increased out of the page, with your thumb pointing up the fingers are coiling counterclockwise. However, Lenz's law tells us the current will oppose the change in flux so the current in the coil will be flowing clockwise.

Transformers

Electromagnetic induction is used in a transformer, a device that magnetically couples two circuits together to allow the transfer of energy between the two circuits without requiring motion. Typically, a transformer consists of a couple of coils and a magnetic core. A changing voltage applied to one coil (the primary) creates a flux in the magnetic core, which induces voltage in the other coil (the secondary). All transformers operate on this simple principle though they range in size and function from those in tiny microphones to those that connect the components of the US power grid.

One of the most important functions of transformers is that they allow us to "step-up" and "step-down" between vastly different voltages. To determine how the voltage is changed by a transformer, we employ any of the following relationships:

$$\frac{V_s}{V_p} = \frac{n_s}{n_p} = \frac{I_p}{I_s}$$

where V_s=secondary voltage
V_p=primary voltage
n_s=number of turns on secondary coil
n_p=number of turns on primary coil
I_p=primary current
I_s=secondary current

Problem: If a step-up transformer has 500 turns on its primary coil and 800 turns on its secondary coil, what will be the output (secondary) voltage be if the primary coil is supplied with 120 V?

Solution:

$$\frac{V_s}{V_p} = \frac{n_s}{n_p}$$

$$V_s = \frac{n_s}{n_p} \times V_p = \frac{800}{500} \times 120V = 192V$$

Motors

Electric motors are found in many common appliances such as fans and washing machines. The operation of a motor is based on the principle that a magnetic field exerts a force on a current carrying conductor. This force is essentially due to the fact that the current carrying conductor itself generates a magnetic field; the basic principle that governs the behavior of an electromagnet. In a motor, this idea is used to convert **electrical energy into mechanical energy**, most commonly rotational energy. Thus the components of the simplest motors must include a strong magnet and a current-carrying coil placed in the magnetic field in such a way that the force on it causes it to rotate.

Motors may be run using DC or AC current and may be designed in a number of ways with varying levels of complexity. A very basic DC motor consists of the following components:
- A **field magnet**
- An **armature** with a coil around it that rotates between the poles of the field magnet
- A **power supply** that supplies current to the armature
- An **axle** that transfers the rotational energy of the armature to the working parts of the motor
- A set of **commutators** and **brushes** that reverse the direction of power flow every half rotation so that the armature continues to rotate

Generators

Generators are devices that are the opposite of motors in that they convert **mechanical energy into electrical energy**. The mechanical energy can come from a variety of sources; combustion engines, blowing wind, falling water, or even a hand crank or bicycle wheel. Most generators rely upon electromagnetic induction to create an electrical current. These generators basically consist of magnets and a coil. The magnets create a magnetic field and the coil is located within this field. Mechanical energy, from whatever source, is used to spin the coil within this field. As stated by Faraday's Law, this produces a voltage.

DOMAIN III. OPTICS AND WAVES

Skill 1   Wave speed, amplitude, wavelength, and frequency

To fully understand waves, it is important to understand many of the terms used to characterize them.

Wave velocity: Two velocities are used to describe waves. The first is phase velocity, which is the rate at which a wave propagates. For instance, if you followed a single crest of a wave, it would appear to move at the phase velocity. The second type of velocity is known as group velocity and is the speed at which variations in the wave's amplitude shape propagate through space. Group velocity is often conceptualized as the velocity at which energy is transmitted by a wave. Phase velocity is denoted v_p and group velocity is denoted v_g. In a medium with refractive index independent of frequency, such as vacuum, the phase velocity is equal to the group velocity.

Crest: The maximum value that a wave assumes; the highest point.

Trough: The lowest value that a wave assumes; the lowest point.

Nodes: The points on a wave with minimal amplitude.

Antinodes: The farthest point from the node on the amplitude axis; both the crests and the troughs are antinodes.

Amplitude: The distance from the wave's highest point (the crest) to the equilibrium point. This is a measure of the maximum disturbance caused by the wave and is typically denoted by A.

Wavelength: The distance between any two sequential troughs or crests denoted λ and representing a complete cycle in the repeated wave pattern.

Period: The time required for a complete wavelength or cycle to pass a given point. The period of a wave is usually denoted T.

Frequency: The number of periods or cycles per unit time (usually a second). The frequency is denoted f and is the inverse of the wave's period (that is, $f=1/T$).

Phase: This is a given position in the cycle of the wave. It is most commonly used in discussing a "being out of phase" or a "phase shift", an offset between waves.

We can visualize several of these terms on the following diagram of a simple, periodic sine wave on a scale of distance displacement (x-axis) vs. (y-axis):

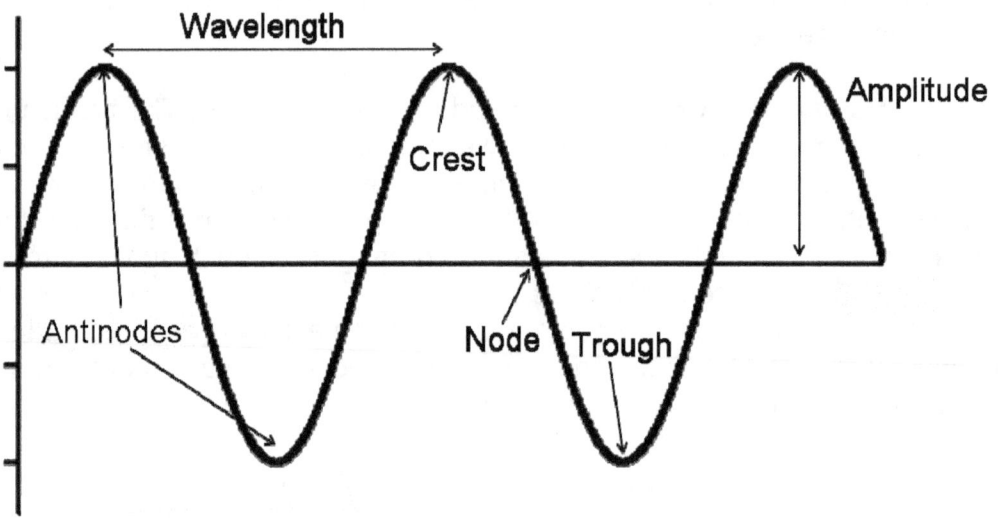

The phase velocity of a wave is related to its wavelength and frequency. Taking light waves, for instance, the speed of light c is equal to the distance traveled divided by time taken. Since the light wave travels the distance of one wavelength λ in the period of the wave T,

$$c = \frac{\lambda}{T}$$

The frequency of a wave, f, is the number of completed periods in one second. In general,

$$f = \frac{1}{T}$$

So the formula for the speed of light can be rewritten as

$$c = \lambda f$$

Thus the phase velocity of a wave is equal to the wavelength times the frequency.

Skill 2   Inverse square law for intensity

For all physical quantities, intensity is a measure of flux over time (see Section VII-2 for a discussion of flux). Light intensity is sometimes referred to as irradiance, radiant emittance, or radiant exitance. The following is the mathematical relationship between intensity and power.

$$I \; \alpha \; \frac{P}{r^2}$$

Where I=intensity
P=power
r=distance from light source

Note that intensity is inversely proportional to squared distance from the source. This means there is a steep drop off of light intensity as one moves away from the light source. For instance, if the distance between a light source and an observer is doubled, the intensity is decreased to $(1/2)^2 = 1/4$ of its original value.

The inverse square law may be understood from purely geometric considerations. Since the light from a point source radiates uniformly in all directions, at a distance r from the source the power is distributed over a sphere with radius r and area proportional to r squared.

Skill 3   Reflection, refraction, absorption, transmission, and scattering (Snell's law, Rayleigh scattering)

Light interacts with matter by **reflection, absorption** or **transmission**. In a transparent material such as a lens, most of the light is transmitted through. Opaque objects such as rocks or cars partially absorb and partially reflect light.

The image you see in a mirrored surface is the result of the reflection of the light waves off the surface. Light waves follow the "law of reflection," i.e. the angle at which the light wave approaches a flat reflecting surface is equal to the angle at which it leaves the surface. Scattering is a form of reflection where the reflection happens in multiple directions. **Rayleigh scattering** is the scattering of an electromagnetic wave by particles that are much smaller than its wavelength. The amount of scattering is inversely related to the wavelength of the wave. Thus, the greater scattering of blue light compared to red light by particles in the atmosphere results in the sky appearing blue.

Absorption is the transfer of energy from a light wave to particles of matter. In the absorption process, a material converts some of the light energy into heat. Some of the energy may be radiated at a different frequency. When light crosses the boundary between two different media, its path is bent, or refracted. For a detailed discussion of **refraction** and **Snell's law** see **section III.8**.

Skill 4 **Transverse and longitudinal waves and their properties (Doppler effect, resonance and natural frequencies, polarization)**

Transverse waves: Waves in which the oscillations are perpendicular to the direction in which in the wave travels.

Longitudinal waves: Waves in which the oscillations are in the direction in which the wave travels.

Polarization: A property of transverse waves that describes the plane perpendicular to the direction of travel in which the oscillation occurs. Note that longitudinal waves are not polarized because they can oscillate only in one direction, the direction of travel. In unpolarized light, the transverse oscillation occurs in all planes perpendicular to the direction of travel. Polarized light (created, for instance, by using polarizing filters that absorb light oscillating in other planes) oscillates in only a selected plane. An everyday example of polarization is found in polarized sunglasses which reduce glare.

Example: A polarizing filter with a horizontal axis will allow the portion of the light waves that are aligned horizontally to pass through the filter and will block the portion of the light waves that are aligned vertically. One-half of the light is being blocked or conversely, one-half of the light is being absorbed. The image being viewed is not distorted but dimmed.

Example: If two filters are used, one with a horizontal axis and one with a vertical axis, all light will be blocked.

The **Doppler effect** is the name given to the perceived change in frequency that occurs when the observer or source of a wave is moving. Specifically, the perceived frequency increases when a wave source and observer move toward each other and decreases when a source and observer move away from each other. Thus, the source and/or observer velocity must be factored in to the calculation of the perceived frequency. The mathematical statement of this effect is:

$$f' = f_0 \left(\frac{v \pm v_o}{v \pm v_s} \right)$$

where f'= observed frequency
f_0= emitted frequency
v= the speed of the waves in the medium
v_s= the velocity of the source (positive in the direction away from the observer)
v_o= the velocity of the observer (positive in the direction towards the source)

Note that any motion that changes the perceived frequency of a wave will cause the Doppler effect to occur. Thus, the wave source, the observer position, or the medium through which the wave travels could possess a velocity that would alter the observed frequency of a wave. You may view animations of stationary and moving wave sources at the following URL:

http://www.kettering.edu/~drussell/Demos/doppler/doppler.html

So, let's consider two examples involving sirens and analyze what happens when either the source or the observer moves. First, imagine a person standing on the side of a road and a police car driving by with its siren blaring. As the car approaches, the velocity of the car will mean sound waves will "hit" the observer as the car comes closer and so the pitch of the sound will be high. As it passes, the pitch will slide down and continue to lower as the car moves away from the observer. This is because the sound waves will "spread out" as the source recedes. Now consider a stationary siren on the top of fire station and a person driving by that station. The same Doppler effect will be observed: the person would hear a high frequency sound as he approached the siren and this frequency would lower as he passed and continued to drive away from the fire station.

The Doppler effect is observed with all types of electromagnetic radiation. In everyday life, we may be most familiar with the Doppler effect and sound, as in the example above. However, we can observe example of it throughout the EM spectrum. For instance, the Doppler effect for light has been exploited by astronomers to measure the speed at which stars and galaxies are approaching. Another familiar application is the use of Doppler radar by police to detect the speed of on coming cars.

All objects vibrate at a particular frequency (or set of frequencies) known as the **natural frequency** that depends on the physical parameters of the object. When energy is fed into an oscillator through an external vibrating source, maximum energy is absorbed when the external source vibrates at the natural frequency of the object. This phenomenon is known as **resonance** and it results in the build up of the amplitude of vibration. It is easy to visualize this by considering a playground swing which is a pendulum with one natural frequency. If you want the swing to go higher and higher you have to match your pushes with the natural frequency of the swing.

Skill 5   Sound waves (pitch and loudness, air columns and standing waves open at both ends and closed at one end, harmonics, beats)

Pitch: The frequency of a sound wave as perceived by the human ear. A high pitch sound corresponds to a high frequency sound wave and a low pitch sound corresponds to a low frequency sound wave.

Sound power: This is the sonic energy of a sound wave per unit time.

Sound intensity: This is simply the sound power per unit area or the loudness of the sound.

A **standing wave** typically results from the interference between two waves of the same frequency traveling in opposite directions. The result is a stationary vibration pattern. One of the key characteristics of standing waves is that there are points in the medium where no movement occurs. The points are called nodes and the points where motion is maximal are called antinodes. This property allows for the analysis of various typical standing waves.

Vibrating string

Imagine a string of length L tied tightly at its two ends. We can generate a standing wave by plucking the string. The waves traveling along the string are reflected at the fixed end points and interfere with each other to produce standing waves. There will always be two nodes at the ends where the string is tied. Depending on the frequency of the wave that is generated, there may also be other nodes along the length of the string. In the diagrams below, examples are given of strings with 0, 1, or 2 additional nodes. These vibrations are known as the 1st, 2nd, and 3rd **harmonics**. The higher order harmonics follow the same pattern although they are not diagramed here.

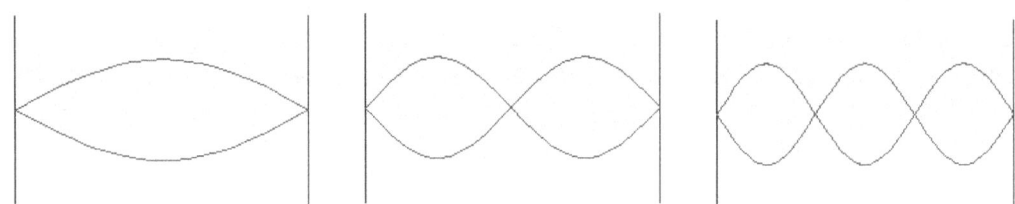

Since we know the length of the string (L) in each case, we can calculate the wavelength (λ) and frequency (f) for any harmonic using the following formula, where n=the harmonic order (n=1,2,3...) and v is the phase velocity of the wave.

$$\lambda_n = \frac{2L}{n} \qquad f_n = \frac{v}{\lambda_n} = n\frac{v}{2L}$$

Waves in a tube

Just as on a string, standing waves can propagate in gaseous or liquid medium inside a tube. In these cases we will also observe harmonic vibrations, but their nature will depend on whether the ends of the tube are closed or open. Specifically, an antinode will be observed at an open end and a node will appear at a closed end. Then, just as in the string example above, we can derive formulas that allow us to predict the wavelength and frequency of the harmonic vibrations that occur in a tube. Below, only frequencies are given; wavelength can be found by applying the formula f=v/λ.

For a tube with two closed ends or two open ends (note that this is the same as for the string described above):

$$f_n = \frac{nv}{2L}$$

where n = 1,2,3,4…

When only one end of a tube is closed, that end become a node and wave exhibits odd harmonics. For a tube with one close end and one open end:

$$f_n = \frac{nv}{4L}$$

where n = 1,3,5,7…

An animation at the following URL may be helpful in visualizing these various standing waves:

http://www.physics.smu.edu/~olness/www/05fall1320/applet/pipe-waves.html

When two sound waves with slightly different frequencies interfere with each other, **beats** result. We hear a beat as a periodic variation in volume with a rate that depends on the difference between the two frequencies. You may have observed this phenomenon when listening to two instruments being tuned to match; beating will be heard as the two instruments approach the same note and disappear when they are perfectly in tune.

Skill 6   Electromagnetic spectrum (frequency regions, color)

The electromagnetic spectrum is measured using frequency (f) in hertz or wavelength (λ) in meters. The frequency times the wavelength of every electromagnetic wave equals the speed of light (3.0×10^8 meters/second).

Roughly, the range of wavelengths of the electromagnetic spectrum is:

	f	**λ**
Radio waves	$10^5 - 10^{-1}$ hertz	$10^3 - 10^9$ meters
Microwaves	$3 \times 10^9 - 3 \times 10^{11}$ hertz	$10^{-3} - 10^{-1}$ meters
Infrared radiation	$3 \times 10^{11} - 4 \times 10^{14}$ hertz	$7 \times 10^{-7} - 10^{-3}$ meters
Visible light	$4 \times 10^{14} - 7.5 \times 10^{14}$ hertz	$4 \times 10^{-7} - 7 \times 10^{-7}$ meters
Ultraviolet radiation	$7.5 \times 10^{14} - 3 \times 10^{16}$ hertz	$10^{-8} - 4 \times 10^{-7}$ meters
X-Rays	$3 \times 10^{16} - 3 \times 10^{19}$ hertz	$10^{-11} - 10^{-8}$ meters
Gamma Rays	$> 3 \times 10^{19}$ hertz	$< 10^{-11}$ meters

Radio waves are used for transmitting data. Common examples are television, cell phones, and wireless computer networks. Microwaves are used to heat food and deliver Wi-Fi service. Infrared waves are utilized in night vision goggles. Visible light we are all familiar with as the human eye is most sensitive to this wavelength range. Light of different colors have different wavelengths. In the visible range, red light has the largest wavelength while violet light has the smallest. UV light causes sunburns and would be even more harmful if most of it were not captured in the Earth's ozone layer. X-rays aid us in the medical field and gamma rays are most useful in the field of astronomy.

Skill 7   **Principle of linear superposition and interference (diffraction, dispersion, beats and standing waves, interference in thin films and Young's double-slit experiment)**

According to the **principle of linear superposition**, when two or more waves exist in the same place, the resultant wave is the sum of all the waves, i.e. the amplitude of the resulting wave at a point in space is the sum of the amplitudes of each of the component waves at that point.

Interference occurs when two or more waves are superimposed. Usually, interference is observed in coherent waves, well-correlated waves that have very similar frequencies or even come from the same source. Superposition of waves may result in either constructive or destructive interference. Constructive interference occurs when the crests of the two waves meet at the same point in time. Conversely, destructive interference occurs when the crest of one wave and the trough of the other meet at the same point in time. It follows, then, that constructive interference increases amplitude and destructive interference decreases it. We can also consider interference in terms of wave phase; waves that are out of phase with one another will interfere destructively while waves that are in phase with one another will interfere constructively. In the case of two simple sine waves with identical amplitudes, for instance, amplitude will double if the waves are exactly in phase and drop to zero if the waves are exactly 180° out of phase.

Additionally, interference can create a **standing wave**, a wave in which certain points always have amplitude of zero. Thus, the wave remains in a constant position. Standing waves typically results when two waves of the same frequency traveling in opposite directions through a single medium are superposed. View an animation of how interference can create a standing wave at the following URL:

http://www.glenbrook.k12.il.us/GBSSCI/PHYS/mmedia/waves/swf.html

For more information about standing waves and **beats** see **section III.5**.

All wavelengths in the EM spectrum can experience interference but it is easy to comprehend instances of interference in the spectrum of visible light. One classic example of this is **Thomas Young's double-slit experiment**. In this experiment a beam of light is shone through a paper with two slits and a striated wave pattern results on the screen. The light and dark bands correspond to the areas in which the light from the two slits has constructively (bright band) and destructively (dark band) interfered. Light from any source can be used to obtain interference patterns. For example, Newton's rings can be produced with sun light. However, in general, white light is less suited for producing clear interference patterns as it is a mix of a full spectrum of colors. Sodium light is close to monochromatic and is thus more suitable for producing interference patterns. The most suitable is laser light as it is almost perfectly monochromatic.

Problem: The interference maxima (location of bright spots created by constructive interference) for double-slit interference are given by

$$\frac{n\lambda}{d} = \frac{x}{D} = \sin\theta \quad n=1,2,3...$$

where λ is the wavelength of the light, d is the distance between the two slits, D is the distance between the slits and the screen on which the pattern is observed and x is the location of the nth maximum. If the two slits are 0.1mm apart, the screen is 5m away from the slits, and the first maximum beyond the center one is 2.0 cm from the center of the screen, what is the wavelength of the light?

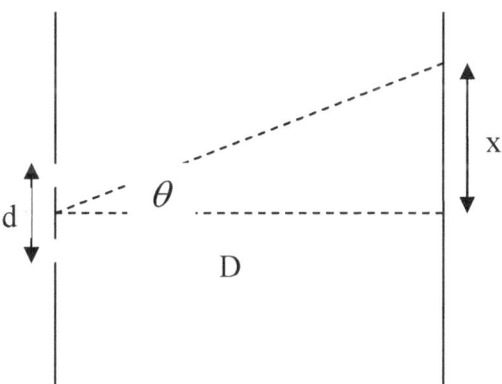

Solution: λ = xd/(Dn) = 0.02 x 0.0001/ (5 x1) = 400 nanometers

Thin-film interference occurs when light waves reflecting off the top surface of a film interfere with the waves reflecting from the bottom surface. We see colors in soap bubbles or in oil films floating on water since the criteria for constructive or destructive interference depend on the wavelength of the light. Non-reflective coatings on materials make use of destructive thin-film interference.

Diffraction is an important characteristic of waves. This occurs when part of a wave front is obstructed. Diffraction and interference are essentially the same physical process. Diffraction refers to various phenomena associated with wave propagation such as the bending, spreading, and interference of waves emerging from an aperture. It occurs with any type of wave including sound waves, water waves, and electromagnetic waves such as light and radio waves.

Here, we take a close look at important phenomena like single-slit diffraction, double-slit diffraction, diffraction grating, other forms of diffraction and lastly interference.

1. Single-slit diffraction: The simplest example of diffraction is single-slit diffraction in which the slit is narrow and a pattern of semi-circular ripples is formed after the wave passes through the slit.

2. Double-slit diffraction: These patterns are formed by the interference of light diffracting through two narrow slits.

3. Diffraction grating: Diffraction grating is a reflecting or transparent element whose optical properties are periodically modulated. In simple terms, diffraction gratings are fine parallel and equally spaced grooves or rulings on a material surface. When light is incident on a diffraction grating, light is reflected or transmitted in discrete directions, called diffraction orders. Because of their light dispersive properties, gratings are commonly used in monochromators and spectrophotometers. Gratings are usually designated by their groove density, expressed in grooves/millimeter. A fundamental property of gratings is that the angle of deviation of all but one of the diffracted beams depends on the wavelength of the incident light.

4. Other forms of diffraction:
i) Particle diffraction: It is the diffraction of particles such as electrons, which is used as a powerful argument for quantum theory. It is possible to observe the diffraction of particles such as neutrons or electrons and hence we are able to infer the existence of wave particle duality.
ii) Bragg diffraction: This is diffraction from a multiple slits, and is similar to what occurs when waves are scattered from a periodic structure such as atoms in a crystal or rulings on a diffraction grating. Bragg diffraction is used in X-ray crystallography to deduce the structure of a crystal from the angles at which the X-rays are diffracted from it.

Dispersion is the separation of a wave into its constituent wavelengths due to interaction with a material occurring in a wavelength-dependent manner (as in thin-film interference for instance). For dispersion of light through a **prism** see **section III.11**.

Skill 8   **Reflection and refraction (Snell's law, total internal reflection, fiber optics)**

Wave **refraction** is a change in direction of a wave due to a change in its speed. This most commonly occurs when a wave passes from one material to another, such as a light ray passing from air into water or glass. However, light is only one example of refraction; any type of wave can undergo refraction. Another example would be physical waves passing from water into oil. At the boundary of the two media, the wave velocity is altered, the direction changes, and the wavelength increases or decreases. However, the frequency remains constant.

Snell's Law describes how light bends, or refracts, when traveling from one medium to the next. It is expressed as

$$n_1 \sin\theta_1 = n_2 \sin\theta_2$$

where n_i represents the index of refraction in medium i, and θ_i represents the angle the light makes with the normal in medium i.

Problem: The index of refraction for light traveling from air into an optical fiber is 1.44. (a) In which direction does the light bend? (b) What is the angle of refraction inside the fiber, if the angle of incidence on the end of the fiber is 22°?

Solution: (a) The light will bend toward the normal since it is traveling from a rarer region (lower n) to a denser region (higher n).

(b) Let air be medium 1 and the optical fiber be medium 2:

$$n_1 \sin\theta_1 = n_2 \sin\theta_2$$
$$(1.00)\sin 22° = (1.44)\sin\theta_2$$
$$\sin\theta_2 = \frac{1.00}{1.44}\sin 22° = (.6944)(.3746) = 0.260$$
$$\theta_2 = \sin^{-1}(0.260) = 15°$$

The angle of refraction inside the fiber is $15°$.

Light travels at different speeds in different media. The speed of light in a vacuum is represented by

$$c = 2.99792458 \times 10^8 \, m/s$$

but is usually rounded to

$$c = 3.00 \times 10^8 \, m/s.$$

PHYSICS

Light will never travel faster than this value. The **index of refraction**, n, is the amount by which light slows in a given material and is defined by the formula

$$n = \frac{c}{v}$$

where v represents the speed of light through the given material.

Problem: The speed of light in an unknown medium is measured to be $1.24 \times 10^8 \, m/s$. What is the index of refraction of the medium?

Solution:

$$n = \frac{c}{v}$$

$$n = \frac{3.00 \times 10^8}{1.24 \times 10^8} = 2.42$$

Referring to a standard table showing indices of refraction, we would see that this index corresponds to the index of refraction for diamond.

Reflection is the change in direction of a wave at an interface between two dissimilar media such that the wave returns into the medium from which it originated. The most common example of this is light waves reflecting from a mirror, but sound and water waves can also be reflected. The law of reflection states that the angle of incidence is equal to the angle of reflection.

Reflection may occur whenever a wave travels from a medium of a given refractive index to another medium with a different index. A certain fraction of the light is reflected from the interface and the remainder is refracted. However, when the wave is moving from a dense medium into one less dense, that is the refractive index of the first is greater than the second, a critical angle exists which will create a phenomenon known as **total internal reflection**. In this situation all of the wave incident at an angle greater than the critical angle is reflected. When a wave reflects off a more dense material (higher refractive index) than that from which it originated, it undergoes a 180° phase change. In contrast, a less dense, lower refractive index material will reflect light in phase.

Fiber optics makes use of the phenomenon of total internal reflection. The light traveling through a fiber reflects off the walls at angles greater than the critical angle and thus keeps the wave confined to the narrow fiber.

Skill 9   Thin lenses

A lens is a device that causes electromagnetic radiation to converge or diverge. The most familiar lenses are made of glass or plastic and designed to concentrate or disperse visible light. Two of the most important parameters for a lens are its thickness and it's focal length. Focal length is a measure of how strongly light is concentrated or dispersed by a lens. For a convex or converging lens, the focal length is the distance at which a beam of light will be focused to a single spot. Conversely, for a concave or diverging lens, the focal length is the distance to the point from which a beam appears to be diverging.

A thin lens in one in which focal length is much greater than lens thickness. For problems involving thin lenses, we can disregard any optical effects of the lens itself. Additionally, we can assume that the light that interacts with the lens makes a small angle with the optical axis of the system and so the sine and tangent values of the angle are approximately equal to the angle itself. This paraxial approximation, along with the thin lens assumptions, allows us to state:

$$\frac{1}{s} + \frac{1}{s'} = \frac{1}{f}$$

> Where s=distance from the lens to the object (object location)
> s'=distance from the lens to the image (image location)
> f=focal length of the lens

Most lenses also cause some magnification of the object. Magnification is defined as:

$$m = \frac{y'}{y} = -\frac{s'}{s}$$

> Where m=magnification
> y'=image height
> y=object height

The images produced by lenses can be either virtual or real. A virtual image is one that is created by rays of light that appear to diverge from a certain point. Virtual images cannot be seen on a screen because the light rays do not actually meet at the point where the image is located. If an image and object appear on the same side of a converging lens, that image is defined as virtual. For virtual images, the image location will be negative and the magnification positive. Real images, on the other hand, are formed by light rays actually passing through the image. Thus, real images are visible on a screen. Real images created by a converging lens are inverted and have a positive image location and negative magnification.

Sign conventions will make it easier to understand thin lens problems:

Focal length: positive for a converging lens; negative for a diverging lens
Object location: positive when in front of the lens; negative when behind the lens
Image location: positive when behind the lens; negative when in front of the lens
Image height: positive when upright; negative when upside-down.
Magnification: positive for an erect, virtual image; negative for an inverted, real image

Example:
A converging lens has a focal length of 10.00 cm and forms a 2.0 cm tall image of a 4.00 mm tall real object to the left of the lens. If the image is erect, is the image real or virtual? What are the locations of the object and the image?

Solution:

We begin by determining magnification:

$$m = \frac{y'}{y} = \frac{0.02m}{0.004m} = 5$$

Since the magnification is positive and the image is erect, we know the image must be virtual.

To find the locations of the object and image, we first relate them by using the magnification:

$$m = -\frac{s'}{s}$$
$$s' = -ms$$

Then we substitute into the thin lens equation, creating one variable in one unknown:

$$\frac{1}{s} + \frac{1}{s'} = \frac{1}{f}$$

$$\frac{1}{s} - \frac{1}{5s} = \frac{1}{10cm}$$

$$\frac{5-1}{5s} = \frac{1}{10cm}$$

$$s = \frac{40cm}{5} = 8cm \rightarrow \rightarrow s' = -5 \times 8cm = -40cm$$

Thus the object is located 8 cm to the left of the lens and the image is 40 cm to the left of the lens.

Skill 10 **Plane and spherical mirrors**

Plane mirrors

Plane mirrors form virtual images. In other words, the image is formed behind the mirror where light does not actually reach. The image size is equal to the object size and object distance is equal to the image distance; i.e. the image is the same distance behind the mirror as the object is in front of the mirror. Another characteristic of plane mirrors is left-right reversal.

Example: Suppose you are standing in front of a mirror with your right hand raised. The image in the mirror will be raising its left hand.

Problem: If a cat creeps toward a mirror at a rate of 0.20 m/s, at what speed will the cat and the cat's image approach each other?

Solution: In one second, the cat will be 0.20 meters closer to the mirror. At the same time, the cat's image will be 0.20 meters closer to the cat. Therefore, the cat and its image are approaching each other at the speed of 0.40 m/s.

Problem: If an object that is two feet tall is placed in front of a plane mirror, how tall will the image of the object be?

Solution: The image of the object will have the same dimensions as the actual object, in this case, a height of two feet. This is because the magnification of an image in a plane mirror is 1.

Curved mirrors

Curved mirrors are usually sections of spheres. In a concave mirror the inside of the spherical surface is silvered while in a convex mirror it is the outside of the spherical surface that is silvered.

Terminology associated with spherical mirrors:

Principal axis: The line joining the center of the sphere (of which we imagine the mirror is a section) to the center of the reflecting surface.

Center of curvature: The center of the sphere of which the mirror is a section.

Vertex: The point on the mirror where the principal axis meets the mirror or the geometric center of the mirror.

PHYSICS

Focal point: The point at which light rays traveling parallel to the principal axis will meet after reflection in a concave mirror. For a convex mirror, it is the point from which light rays traveling parallel to the principal axis will appear to diverge from after reflection. The focal point is midway between the center of curvature and the vertex.

Focal length: The distance between the focal point and the vertex.

Radius of curvature: The distance between the center of curvature and the vertex, i.e. the radius of the sphere of which the mirror is a section. The radius of curvature is twice the focal length.

The relationship between the object distance from vertex s, the image distance from vertex s', and the focal length f is given by the equation

$$\frac{1}{s} + \frac{1}{s'} = \frac{1}{f}$$

Magnification is defined as:

$$m = \frac{y'}{y} = -\frac{s'}{s}$$

where m=magnification, y'=image height, y=object height

Image characteristics for **concave mirrors**:
1) If the object is located **beyond the center of curvature**, the image will be **real, inverted, smaller** and located between the focal point and center of curvature.
2) If the object is located **at the center of curvature**, the image will be **real, inverted, of the same height** and also located at the center of curvature.
3) If the object is located **between the center of curvature and focal point**, the image will be **real, inverted, larger** and located beyond the center of curvature.
4) If the object is located **at the focal point no image is formed**.
5) If the object is located **between the focal point and vertex**, the image will be **virtual, upright, larger** and located on the opposite side of the mirror.
6) If the object is located **at infinity** (very far away), the image is **real, inverted, smaller** and located at the focal point.

For **convex mirrors**, the image is always **virtual, upright, reduced in size** and formed on the opposite side of the mirror.

Problem: A concave mirror collects light from a star. If the light rays converge at 50 cm, what is the radius of curvature of the mirror?

Solution: The focal length, in this case, 50 cm, is the distance from the focal point to the mirror. Since the focal point is the midpoint of the line from the vertex to the center of curvature, or focal length, the focal length would be one-half the radius of curvature. Since the focal length in this case is 50 cm, the radius of curvature would be 100 cm.

Problem: An image of an object in a mirror is upright and reduced in size. In what type of mirror is this image being viewed, plane, concave, or convex?

Solution: The image in a plane mirror would be the same size as the object. The image in a concave mirror would be magnified if upright. Only a convex mirror would produce a reduced upright image of an object.

Skill 11   Prisms

Prisms are transparent devices used in optics to manipulate electromagnetic waves, especially light. Any largely transparent material can be used to make a prism, but glass, plastics, and calcite crystals are among the most common. Prisms rely on the fact that light changes speed as it moves from one medium to another. This then causes the light to be bent and/or reflected. The degree to which bending or reflection occurs is a function of the light's angle of incidence and the refractive indices of the media.

Prisms can perform various operations on EM waves and are classified according these actions. The three main types of prisms are:

Dispersive prisms: This type of prism breaks light up into its spectral components (i.e., blue light, yellow light, red light, etc). This class includes the familiar triangular prisms, through which you may have seen sunlight separated into an apparent "rainbow". Dispersive prisms separate white light into these constituent colors by relying on the differences in refractive index that result from the varying frequencies of the light.

Polarizing prisms: These prisms are sometimes known simply as polarizers. They convert nonpolarized light (or other types of EM waves) into beams with a single polarization state. (See section III.4 for description of polarization). Cameras and many optical instruments contain polarizing prisms.

Reflective prisms: These prisms reflect light and are used in constructing binoculars. In certain cases, total internal reflection occurs and the prism may substitute for a mirror.

Skill 12   Optical instruments (simple magnifier, microscope, telescope)

Simple magnifier
A simple magnifier is a convex or converging lens that allows a user to place an object closer to the eye than the near point (the distance within which objects become blurry, assumed to be approximately 25 cm) and view an enlarged virtual image. The magnification achieved is given by 25/f where f is the focal length of the magnifying lens.

Telescope
A telescope is a device that has the ability to make distant objects appear to be much closer. Most telescopes are one of two varieties, a refractor which uses lenses or a reflector which uses mirrors. Each accomplishes the same purpose but in totally different ways. The basic idea of a telescope is to collect as much light as possible, focus it, and then magnify it. The objective lens or primary mirror of a telescope brings the light from an object into focus. An eyepiece lens takes the focused light and "spreads it out" or magnifies it using the same principle as a magnifying glass using two curved surfaces to refract the light.

Microscope
Microscopes are used to view objects that are too small to be seen with the naked eye. A microscope usually has an objective lens that collects light from the sample and an eyepiece which brings the image into focus for the observer. It also has a light source to illuminate the sample. Typical optical microscopes achieve magnification of up to 1500 times.

Eye
The eye is a very complex sensory organ. Although there are many critical anatomical features of the eye, the lens and retina are the most important for focusing and sensing light. Light passes through the cornea and through the lens. The lens is attached to numerous muscles that contract and relax to move the lens in order to focus the light onto the retina. The pupil also contracts and relaxes to allow more or less light in the eye as required. The eye relies on refraction to focus light onto the retina. Refraction occurs at four curved interfaces; between the air and the front of the cornea, the back of the cornea and the aqueous humor, the aqueous humor at the front of the lens, and the back of the lens and the vitreous humor. When each of these interfaces are working properly the light arrives at the retina in perfect focus for transmission to the brain as an image.

Eye Glasses

When all the parts of the eye are not working together correctly, corrective lenses or eyeglasses may be needed to assist the eye in focusing the light onto the retina. The surfaces of the lens or cornea may not be smooth causing the light to refract in the wrong direction. This is called astigmatism. Another common problem is that the lens is not able to change its curvature appropriately to match the image. The cornea can also be misshaped resulting in blurred vision. Corrective lenses consist of curves pieces of glass which bend the light in order to change the focal point of the light. A nearsighted eye forms images in front of the retina. To correct this, a minus lens consisting of two concave prisms is used to bend light out and move the image back to the retina. A farsighted eye creates images behind the retina. This is corrected using plus lenses that bend light in and bring the image forward onto the retina. The worse the vision, the farther out of focus the image is on the retina. Therefore the stronger the lens the further the focal point is moved to compensate.

Spectroscope

Spectrometers known as spectroscopes are used to identify materials. Spectroscopes are used often in astronomy and some branches of chemistry. Early spectroscopes were simply a prism with graduations marking wavelengths of light. Modern spectroscopes typically use a diffraction grating, a movable slit, and some kind of photo detector, all automated and controlled by a computer. When materials are heated they emit light that is characteristic of its atomic composition. The emission of certain frequencies of light produce a pattern of lines that are comparable to a fingerprint. The yellow light emission of heated sodium is a typical example.

A spectroscope is able to detect, measure and record the frequencies of the emitted light. This is done by passing the light though a slit to a collimating lens which transforms the light into parallel rays. The light is then passed through a prism that refracts the beam into a spectrum of different wavelengths. The image is then viewed alongside a scale to determine the characteristic wavelengths. Spectral analysis is an important tool for determining and analyzing the composition of unknown materials as well as for astronomical studies.

Camera

A camera is another device that utilizes the lens' ability to refract light to capture and process an image. As with the eye, light enters the lens of a camera and focuses the light on the other side. Instead of focusing on the retina, the image is focused on the film to create a film negative. This film negative is later processed with chemicals to create a photograph. A camera uses a converging or convex lens. This lens captures and directs light to a single point to create a real image on the surface of the film. To focus a camera on an image, the distance of the lens from the film is adjusted in order to ensure that the real image converges on the surface of the film and not in front of or behind it.

Different lenses are available which capture and bend the light to different degrees. A lens with more pronounced curvature will be able to bend the light more acutely causing the image to converge more closely to the lens. Conversely a flatter lens will have a longer focal distance. The further the lens is located from the film (flatter lens), the larger the image becomes. Thus zoom lenses on cameras are flat while wide angle lenses are more rounded. The focal length number on a certain lens conveys the magnification ability of the lens.

The film functions like the retina of the eye in that it is light sensitive and can capture light images when exposed. However, this exposure must be brief to capture the contrasting amounts of light and a clear image. The rest of the camera functions to precisely control how much light contacts the film. The aperture is the lens opening which can open and close to let in more or less light. The temporal length of light exposure is controlled by the shutter which can be set at different speeds depending on the amount of action and level of light available. The film speed refers to the size of the light sensitive grains on the surface of the film. The larger grains absorb more light photons than the smaller grains, so film speed should be selected according to lighting conditions.

DOMAIN IV.   HEAT AND THERMODYNAMICS

Skill 1   Measurement of heat and temperature (temperature scales)

Heat is the thermal energy a body has due to the kinetic and potential energy of its atoms and molecules and is measured in the same units as any other form of energy, the SI unit being Joule. The traditional unit for the measurement of heat is the calorie that is related to the Joule through the relationship **1 calorie = 4.184 Joule**. Many other forms of energy, mechanical energy when you rub your hands together or electrical energy from a light bulb for instance, can be converted into heat energy. For a detailed discussion of heat energy see section IV.4.

Heat is generally measured in terms of temperature, a measure of the average internal energy of a material. Temperature is an **intensive property**, meaning that it does not depend on the amount of material. Heat content is an **extensive property** because more material at the same temperature will contain more heat. The relationship between the change in heat energy of a material and the change in its temperature is given by $\Delta Q = mC\Delta T$, where ΔQ is the change in heat energy, m is the mass of the material, ΔT is the change in temperature and C is the specific heat which is characteristic of a particular material. Section IV.4 provides a detailed discussion of heat, temperature and specific heat.

There are four generally recognized temperature scales, Celsius, Fahrenheit, Kelvin and Rankine. The Kelvin and Rankine scales are absolute temperature scales corresponding to the Celsius and Fahrenheit scales, respectively. Absolute temperature scales have a zero reading when the temperature reaches absolute zero (the theoretical point at which no thermal energy exists). The absolute temperature scales are useful for many calculations in chemistry and physics.

To convert between Celsius and Fahrenheit, use the following relationship:

$$x \, °F = (5/9)(x - 32) \, °C$$

To convert to the absolute temperature scales, use the appropriate conversion below:

$$x \, °F = x + 459.67 \, °R$$

$$x \, °C = x + 273.15 \, K$$

Note that the size of each degree on the Fahrenheit/ Rankine scale is smaller than the size of a degree on the Celsius/Kelvin scale.

Skill 2   Thermal expansion

Most solid and liquid materials expand when heated with a change in dimension proportional to the change in temperature. A notable exception to this is water between $0^0 C$ and $4^0 C$.

If we consider a long rod of length L that increases in length by ΔL when heated, the fractional change in length $\Delta L / L$ is directly proportional to the change in temperature ΔT.

$$\Delta L / L = \alpha . \Delta T$$

The constant of proportionality α is known as the **coefficient of linear expansion** and is a property of the material of which the rod is made.

Problem: The temperature of an iron rod 10 meters long changes from $-3^0 C$ to $12^0 C$. If iron has a coefficient of linear expansion of 0.000011 per $^0 C$, by how much does the rod expand?

Solution: The length of the rod L = 10 meters.
Change in temperature $\Delta T = 12^0 C - (-3^0 C) = 15^0 C$
Change in length of the rod $\Delta L = 0.000011 \times 10 \times 15 = .00165$ meters

If instead of a rod, we consider an area A that increases by ΔA when heated, we find that the fractional change in area is proportional to the change in temperature ΔT. The proportionality constant in this case is known as the **coefficient of area expansion** and is related to the coefficient of linear expansion as demonstrated below.

If A is a rectangle with dimensions L_1 and L_2, then

$$A + \Delta A = (L_1 + \Delta L_1)(L_2 + \Delta L_2)$$
$$= (L_1 + \alpha L_1 \Delta T)(L_2 + \alpha L_2 \Delta T)$$
$$= L_1 L_2 + 2\alpha L_1 L_2 \Delta T + \alpha^2 (\Delta T)^2$$

Ignoring the higher order term for small changes in temperature, we find that

$$\Delta A = 2\alpha A \Delta T = \gamma A \Delta T$$

Thus the coefficient of area expansion $\gamma = 2\alpha$.

Following the same procedure as above, we can show that the change in volume of a material when heated may be expressed as

$$\Delta V = 3\alpha V \Delta T = \beta V \Delta T$$

where the **coefficient of volume expansion** $\beta = 3\alpha$.

Problem: An aluminum sphere of radius 10cm is heated from $0°C$ to $25°C$. What is the change in its volume? The coefficient of linear expansion of aluminum is 0.000024 per $°C$.

Solution: Volume V of the sphere = $\frac{4}{3}\Pi r^3 = \frac{4}{3} \times 3.14 \times 1000 cm^3 = 4186.67 cm^3$

Change in volume of the sphere = $3 \times 0.000024 \times 4186.67 \times 25 = 7.54 cm^3$

Skill 3 Thermocouples

A thermocouple is a widely used type of temperature sensing device. Thermocouple operation is based on the thermoelectric or Seebeck effect. Put simply, when any conductive material is subjected to the thermal gradient it will generate a voltage. To measure this voltage another conductor must be attached to it. This conductor also experiences a gradient and produces its own voltage. Every conductor (metal) produces a different voltage when exposed to different temperature gradients. By using two different metals to construct the thermocouple circuit, two different and opposing voltages will be generated, which will leave a residual voltage that can be measured to determine the temperature.

An important aspect to remember when using a thermocouple is that absolute temperature cannot be measured, only the temperature gradient which the thermocouple experiences. Typically, the "cold" end of the thermocouple would be held at a known temperature, in order for the hot temperature to be calculated by the thermocouple. The voltage difference can typically be between 1 and about 70 microvolts per degree for the metal combinations typically offered in thermocouples today. One good general purpose thermocouple metal combination is Chromel (Ni-Cr alloy) and Alumel (Ni-Al alloy). Another common thermocouple is composed of copper and Constantan (Cu-Ni alloy).

The main advantages to thermocouples are that they are cheap and interchangeable due to standard connectors. They are also available to measure a wide range of temperatures. However, they are limited in precision and errors in a thermocouple system are typically 1° C or more.

Skill 4   **Heat capacity and specific heat, Latent heat of phase change (heat of fusion, heat of vaporization)**

The **internal energy** of a material is the **sum of the total kinetic energy** of its molecules and the **potential energy** of interactions between those molecules. Total kinetic energy includes the contributions from translational motion and other components of motion such as rotation. The potential energy includes **energy stored in the form of resisting intermolecular attractions** between molecules.

The **enthalpy** (H) of a material is the **sum of its internal energy and the mechanical work** it can do by driving a piston. A change in the **enthalpy** of a substance is the total **energy** change caused by **adding/removing heat** at constant pressure.

When a material is heated and experiences a phase change, **thermal energy is used to break the intermolecular bonds** holding the material together. Similarly, bonds are formed with the release of thermal energy when a material changes its phase during cooling. Therefore, **the energy of a material increases during a phase change that requires heat and decreases during a phase change that releases heat**. For example, the energy of H_2O increases when ice melts and decreases when water freezes.

Heat capacity and specific heat

A substance's molar **heat capacity** is the heat required to **change the temperature of one mole of the substance by one degree**. Heat capacity has units of joules per mol- kelvin or joules per mol- °C. The two units are interchangeable because we are only concerned with differences between one temperature and another. A Kelvin degree and a Celsius degree are the same size.

The **specific heat** of a substance (also called specific heat capacity) is the heat required to **change the temperature of one gram or kilogram by one degree**. Specific heat has units of joules per gram-°C or joules per kilogram-°C.

These terms are used to solve problems involving a change in temperature by applying the formula:

$q = n \times C \times \Delta T$ where $q \Rightarrow$ heat added (positive) or evolved (negative)

        $n \Rightarrow$ amount of material

        $C \Rightarrow$ molar heat capacity if n is in moles, specific heat if n is a mass

        $\Delta T \Rightarrow$ change in temperature $T_{final} - T_{initial}$

Example:
What is the change in energy of 10 g of gold at 25 °C when it is heated beyond its melting point to 1300 °C. You will need the following data for gold:

$$\text{Solid heat capacity: 28 J/mol-K}$$
$$\text{Molten heat capacity: 20 J/mol-K}$$
$$\text{Enthalpy of fusion: 12.6 kJ/mol}$$
$$\text{Melting point: 1064 °C}$$

Solution: First determine the number of moles used: $10 \text{ g} \times \frac{1 \text{ mol}}{197 \text{ g}} = 0.051 \text{ mol}$.

There are then three steps. 1) Heat the solid. 2) Melt the solid. 3) Heat the liquid. All three require energy so they will be positive numbers.

1) Heat the solid:
$$q_1 = n \times C \times \Delta T = 0.051 \text{ mol} \times 28 \frac{J}{\text{mol-K}} \times (1064 \text{ °C} - 25 \text{ °C})$$
$$= 1.48 \times 10^3 \text{ J} = 1.48 \text{ kJ}$$

2) Melt the solid: $q_2 = n \times \Delta H_{fusion} = 0.051 \text{ mol} \times 12.6 \frac{kJ}{mol}$
$$= 0.64 \text{ kJ}$$

3) Heat the liquid:
$$q_3 = n \times C \times \Delta T = 0.051 \text{ mol} \times 20 \frac{J}{\text{mol-K}} \times (1300 \text{ °C} - 1064 \text{ °C})$$
$$= 2.4 \times 10^2 \text{ J} = 0.24 \text{ kJ}$$

The sum of the three processes is the total change in energy of the gold:
$$q = q_1 + q_2 + q_3 = 1.48 \text{ kJ} + 0.64 \text{ kJ} + 0.24 \text{ kJ} = 2.36 \text{ kJ}$$
$$= 2.4 \text{ kJ}$$

A **temperature vs. heat graph** can demonstrate these processes visually. One can also calculate the specific heat or latent heat of phase change for the material by studying the details of the graph.

Example: The plot below shows heat applied to 1g of ice at -40C. The horizontal parts of the graph show the phase changes where the material absorbs heat but stays at the same temperature. The graph shows that ice melts into water at 0C and the water undergoes a further phase change into steam at 100C.

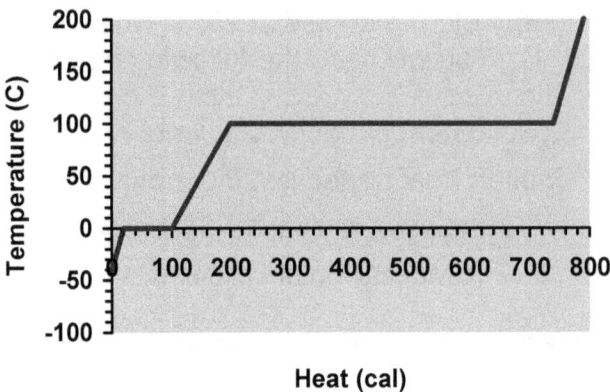

Heat (cal)

The specific heat of ice, water and steam and the latent heat of fusion and vaporization may be calculated from each of the five segments of the graph.

For instance, we see from the flat segment at temperature 0C that the ice absorbs 80 cal of heat. The latent heat L of a material is defined by the equation $\Delta Q = mL$ where ΔQ is the quantity of heat transferred and m is the mass of the material. Since the mass of the material in this example is 1g, the latent heat of fusion of ice is given by $L = \Delta Q / m = 80$ cal/g.

The next segment shows a rise in the temperature of water and may be used to calculate the specific heat C of water defined by $\Delta Q = mC\Delta T$, where ΔQ is the quantity of heat absorbed, m is the mass of the material and ΔT is the change in temperature. According to the graph, ΔQ = 200-100 =100 cal and ΔT = 100-0=100C. Thus, C = 100/100 = 1 cal/gC.

Problem: The plot below shows the change in temperature when heat is transferred to 0.5g of a material. Find the initial specific heat of the material and the latent heat of phase change.

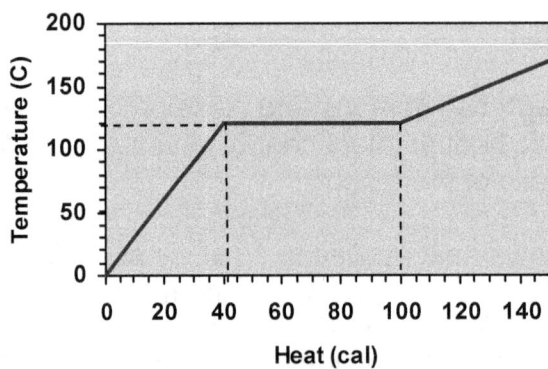

Heat (cal)

Solution: Looking at the first segment of the graph, we see that $\Delta Q = 40$ cal and $\Delta T = 120$ C. Since the mass $m = 0.5$g, the specific heat of the material is given by $C = \Delta Q / (m\Delta T) = 40/(0.5 \times 120) = 0.67$ cal/gC.

The flat segment of the graph represents the phase change. Here $\Delta Q = 100 - 40 = 60$ cal. Thus, the latent heat of phase change is given by $L = \Delta Q / m = 60/(0.5) = 120$ cal/g.

Skill 5   Transfer of heat (conduction, convection and radiation)

All heat transfer is the movement of thermal energy from hot to cold matter. This movement down a thermal gradient is a consequence of the second law of thermodynamics. The three methods of heat transfer are listed and explained below.

Conduction: Electron diffusion or photo vibration is responsible for this mode of heat transfer. The bodies of matter themselves do not move; the heat is transferred because adjacent atoms that vibrate against each other or as electrons flow between atoms. This type of heat transfer is most common when two solids come in direct contact with each other. This is because molecules in a solid are in close contact with one another and so the electrons can flow freely. It stands to reason, then, that metals are good conductors of thermal energy. This is because their metallic bonds allow the freest movement of electrons. Similarly, conduction is better in denser solids. Examples of conduction can be seen in the use of copper to quickly convey heat in cooking pots, the flow of heat from a hot water bottle to a person's body, or the cooling of a warm drink with ice.

The amount of heat transferred by conduction through a material depends on several factors. It is directly proportional to the temperature difference ΔT between the surface from which the heat is flowing and the surface to which it is transferred. Heat flow H increases with the area A through which the flow occurs and also with the time duration t. The thickness of the material reduces the flow of heat. The relationship between all these variables is expressed as

$$H = \frac{k.t.A.\Delta T}{d}$$

where the proportionality constant k is known as the **thermal conductivity**, a property of the material. Thermal conductivity of a good conductor is close to 1 (0.97 cal/cm.s.$^\circ C$ for silver) while good insulators have thermal conductivity that is nearly zero (0.0005 cal/cm.s.$^\circ C$ for wood).

Problem: A glass window pane is 50 cm long and 30 cm wide. The glass is 1 cm thick. If the temperature indoors is $15\,^\circ C$ higher than it is outside, how much heat will be lost through the window in 30 minutes? The thermal conductivity of glass is 0.0025 cal/cm.s.$^\circ C$.

Solution: The window has area A = 1500 sq. cm and thickness d = 1 cm. Duration of heat flow is 1800 s and the temperature difference $\Delta T = 15\,^\circ C$. Therefore heat loss through the window is given by

$$H = (0.0025 \times 1800 \times 1500 \times 15) / 1 = 101250 \text{ calories}$$

Convection: Convection involves some conduction but is distinct in that it involves the movement of warm particles to cooler areas. Convection may be either natural or forced, depending on how the current of warm particles develops. Natural convection occurs when molecules near a heat source absorb thermal energy (typically via conduction), become less dense, and rise. Cooler molecules then take their place and a natural current is formed. Forced convection, as the name suggests, occurs when liquids or gases are moved by pumps, fans, or other means to be brought into contact with warmer or cooler masses. Because the free motion of particles with different thermal energy is key to this mode of heat transfer, convection is most common in liquid and gases. Convection can, however, transfer heat between a liquid or gas and a solid. Forced convection is used in "forced air" home heating systems and is common in industrial manufacturing processes. Additionally, natural convection is responsible for ocean currents and many atmospheric events. Finally, natural convection often arises in association with conduction, for instance in the air near a radiator or the water in a pot on the stove. The mathematical analysis of heat transfer by convection is far more complicated than for conduction or radiation and will not be addressed here.

Radiation: This method of heat transfer occurs via electromagnetic radiation. All matter warmer than absolute zero (that is, all known matter) radiates heat. This radiation occurs regardless of the presence of any medium. Thus, it occurs even in a vacuum. Since light and radiant heat are both part of the EM spectrum, we can easily visualize how heat is transferred via radiation. For instance, just like light, radiant heat is reflected by shiny materials and absorbed by dark materials. Common examples of radiant heat include the way sunlight travels from the sun to warm the earth, the use of radiators in homes, and the warmth of incandescent light bulbs.

The amount of energy radiated by a body at temperature T and having a surface area A is given by the Stefan-Boltzmann law expressed as

$$I = e\sigma A T^4$$

where I is the radiated power in watts, e (a number between 0 and 1) is the **emissivity** of the body and σ is a universal constant known as **Stefan's constant** that has a value of $5.6703 \times 10^{-8} W/m^2.K^4$. Black objects absorb and radiate energy very well and have emissivity close to 1. Shiny objects that reflect energy are not good absorbers or radiators and have emissivity close to zero.

A body not only radiates thermal energy but also absorbs energy from its surroundings. The net power radiation from a body at temperature T in an environment at temperature T_0 is given by

$$I = e\sigma A(T^4 - T_0^4)$$

Problem: Calculate the net power radiated by a body of surface area 2 sq. m, temperature $30°C$ and emissivity 0.5 placed in a room at a temperature of $15°C$.

Solution: $I = 0.5 \times 5.67 \times 10^{-8} \times 2(303^4 - 288^4) = 88$ W

Skill 6 Kinetic molecular theory (ideal gas laws)

The relationship between **kinetic energy** and **intermolecular forces** determines whether a collection of molecules will be a gas, liquid, or solid. In a gas, the energy of intermolecular forces is much weaker than the kinetic energy of the molecules. Kinetic molecular theory is usually applied to gases and is best applied by imagining ourselves shrinking down to become a molecule and picturing what happens when we bump into other molecules and into container walls.

Gas **pressure** results from molecular collisions with container walls. The **number of molecules** striking an **area** on the walls and the **average kinetic energy** per molecule are the only factors that contribute to pressure. A higher **temperature** increases speed and kinetic energy. There are more collisions at higher temperatures, but the average distance between molecules does not change, and thus density does not change in a sealed container.

Kinetic molecular theory explains why the pressure and temperature of **ideal gases** behave the way they do by making a few assumptions, namely:

1) The energies of intermolecular attractive and repulsive forces may be neglected.
2) The average kinetic energy of the molecules is proportional to absolute temperature.
3) Energy can be transferred between molecules during collisions and the collisions are elastic, so the average kinetic energy of the molecules doesn't change due to collisions.
4) The volume of all molecules in a gas is negligible compared to the total volume of the container.

Strictly speaking, molecules also contain some kinetic energy by rotating or experiencing other motions. The motion of a molecule from one place to another is called **translation**. Translational kinetic energy is the form that is transferred by collisions, and kinetic molecular theory ignores other forms of kinetic energy because they are not proportional to temperature.

The following table summarizes the application of kinetic molecular theory to an increase in container volume, number of molecules, and temperature:

	Impact on gas: − = decrease, 0 = no change, + = increase						
Effect of an **increase** in one variable with other two constant	Average distance between molecules	Density in a sealed container	Average speed of molecules	Average translational kinetic energy of molecules	Collisions with container walls per second	Collisions per unit area of wall per second	Pressure (P)
Volume of container (V)	+	−	0	0	−	−	−
Number of molecules	−	+	0	0	+	+	+
Temperature (T)	0	0	+	+	+	+	+

Additional details on the kinetic molecular theory may be found at http://hyperphysics.phy-astr.gsu.edu/hbase/kinetic/ktcon.html. An animation of gas particles colliding is located at http://comp.uark.edu/~jgeabana/mol_dyn/.

The pressure, temperature and volume relationships for an ideal gas (a gas described by the assumptions of the kinetic molecular theory listed above) are given by the following gas laws:

Boyle's law states that the volume of a fixed amount of gas at constant temperature is inversely proportional to the gas pressure, or:

$$V \propto \frac{1}{P}.$$

Gay-Lussac's law states that the pressure of a fixed amount of gas in a fixed volume is proportional to absolute temperature, or:

$$P \propto T.$$

Charles's law states that the volume of a fixed amount of gas at constant pressure is directly proportional to absolute temperature, or:

$$V \propto T.$$

The **combined gas law** uses the above laws to determine a proportionality expression that is used for a constant quantity of gas:

$$V \propto \frac{T}{P}.$$

The combined gas law is often expressed as an equality between identical amounts of an ideal gas at two different states ($n_1 = n_2$):

$$\frac{P_1 V_1}{T_1} = \frac{P_2 V_2}{T_2}.$$

Avogadro's hypothesis states that equal volumes of different gases at the same temperature and pressure contain equal numbers of molecules. **Avogadro's law** states that the volume of a gas at constant temperature and pressure is directly proportional to the quantity of gas, or:

$$V \propto n \text{ where } n \text{ is the number of moles of gas.}$$

Avogadro's law and the combined gas law yield $V \propto \frac{nT}{P}$. The proportionality constant R--the **ideal gas constant**--is used to express this proportionality as the **ideal gas law**:

$$PV = nRT.$$

The ideal gas law is useful because it contains all the information of Charles's, Avogadro's, Boyle's, and the combined gas laws in a single expression.

Solving ideal gas law problems is a straightforward process of algebraic manipulation. **Errors commonly arise from using improper units**, particularly for the ideal gas constant R. An absolute temperature scale must be used—never °C—and is usually reported using the Kelvin scale, but volume and pressure units often vary from problem to problem.

If pressure is given in atmospheres and volume is given in liters, a value for R of **0.08206 L·atm/(mol·K)** is used. If pressure is given in pascal (newtons/m²) and volume in m³, then the SI value for R of **8.314 J/(mol·K)** may be used because a joule is defined as a newton-meter or a pascal·m³. A value for R of **8.314 Pa·m³/(mol·K)** is identical to the ideal gas constant using joules.

The ideal gas law may also be rearranged to determine gas molar density in moles per unit volume (molarity):

$$\frac{n}{V} = \frac{P}{RT}.$$

Gas density d in grams per unit volume is found after multiplication by the molecular weight M:

$$d = \frac{nM}{V} = \frac{PM}{RT}.$$

Molecular weight may also be determined from the density of an ideal gas:

$$M = \frac{dV}{n} = \frac{dRT}{P}.$$

Example: Determine the molecular weight of an ideal gas that has a density of 3.24 g/L at 800 K and 3.00 atm.

Solution: $$M = \frac{dRT}{P} = \frac{\left(3.24 \frac{g}{L}\right)\left(0.08206 \frac{L\text{-atm}}{\text{mol-K}}\right)(800 \text{ K})}{3.00 \text{ atm}} = 70.9 \frac{g}{\text{mol}}.$$

Tutorials for gas laws may be found online at:
http://www.chemistrycoach.com/tutorials-6.htm. A flash animation tutorial for problems involving a piston may be found at
http://www.mhhe.com/physsci/chemistry/essentialchemistry/flash/gasesv6.swf.

Skill 7 First law of thermodynamics (internal energy, energy conservation)

The first law of thermodynamics is a restatement of conservation of energy, i.e. the principle that energy cannot be created or destroyed. It also governs the behavior of a system and its surroundings. The change in heat energy supplied to a system (Q) is equal to the sum of the change in the internal energy (U) and the change in the work (W) done by the system against internal forces.

The internal energy of a material is the sum of the total kinetic energy of its molecules and the potential energy of interactions between those molecules. Total kinetic energy includes the contributions from translational motion and other components of motion such as rotation. The potential energy includes energy stored in the form of resisting intermolecular attractions between molecules. Mathematically, we can express the relationship between the heat supplied to a system, its internal energy and work done by it as

$$\Delta Q = \Delta U + \Delta W$$

Let us examine a sample problem that relies upon this law.

Problem: A closed tank has a volume of 40.0 m³ and is filled with air at 25°C and 100 kPa. We desire to maintain the temperature in the tank constant at 25°C as water is pumped into it. How much heat will have to be removed from the air in the tank to fill the tank ½ full?

Solution: The problem involves isothermal compression of a gas, so $\Delta U_{gas}=0$. Consulting the equation above, $\Delta Q = \Delta U + \Delta W$, it is clear that the heat removed from the gas must be equal to the work done by the gas.

$$Q_{gas} = W_{gas} = P_{gas}V_1 \ln\left(\frac{V_2}{V_T}\right) = P_{gas}V_T \ln\left(\frac{\frac{1}{2}V_T}{V_T}\right) = P_{gas}V_T \ln \frac{1}{2}$$

$$= (100 kPa)(40.0 m^3)(-0.69314) = -2772.58 kJ$$

Thus, the gas in the tank must lose 2772.58 kJ to maintain its temperature.

Skill 8 Thermal processes involving pressure, volume and temperature

For a gas in a state of equilibrium, the pressure P, volume V and temperature T are related through an equation of state. For a **quasi-static process** in which the gas remains close to equilibrium at all times, any two of these variables can characterize the state of the gas at a point in time. On a PV diagram, the state of a gas at any point is represented by its pressure P and volume V. Since P, V and T are related, when the volume changes through expansion or compression, either the pressure or temperature or both of these variables must change.

Processes where pressure remains constant are known as **isobaric** processes. These are represented by horizontal straight lines on a PV diagram.

Isothermal processes are those in which the temperature of the gas remains the same throughout. For an ideal gas $PV = nRT$ = constant. Thus the PV curve is a hyperbola. Since the temperature does not change, the internal energy of the gas remains constant and the heat Q absorbed by the gas is equal to the work W done by the gas.

During an **adiabatic** process no heat flows in or out of the gas. Thus the work done by the gas equals the decrease in internal energy of the gas and the temperature falls as the gas expands. The decrease in temperature leads to a greater decrease in pressure than in the case of isothermal expansion. As a result, the PV curve for adiabatic expansion of a gas is steeper than that for isothermal expansion.

Problem: The PV diagram below shows the quasi-static expansion of a gas from volume V1 to volume V2 through three different paths A, B and C, one adiabatic, one isothermal and one isobaric. Identify the process type of each path.

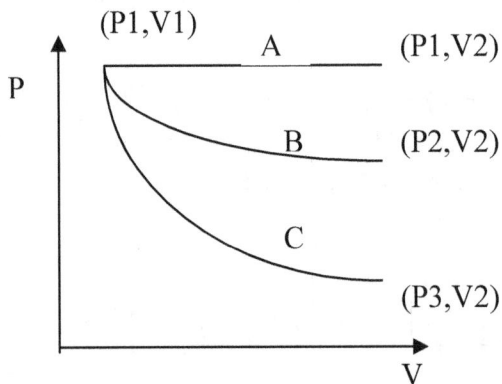

Solution: Path A is isobaric since it is a horizontal line with the pressure remaining constant at P1. Path C is adiabatic and path B is isothermal since adiabatic expansion leads to a steeper fall in pressure.

The work done by the gas or on the gas can be calculated from its *PV* diagram, i.e. the plot of the gas pressure vs. volume.

Consider a gas confined in a cylinder with a frictionless piston. In a quasi-static process, the piston moves very slowly without acceleration. If the area of the piston is *A* and it moves a distance *dx*, then the force exerted by the gas on the piston is *PA* and the work done by the gas is *PAdx*.

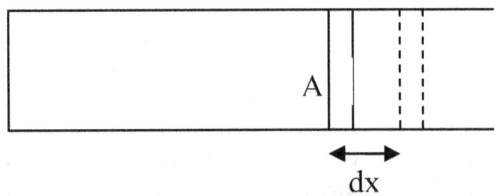

Since the change in volume of the gas *dV* = *Adx*, the work done by the gas is given by *dW* = *PAdx* = *PdV*. If the gas expands quasi-statically from volume *V1* to volume *V2*, the total work done is $W = \int_{V1}^{V2} PdV$. In a *PV* diagram, this expression represents the area under the pressure vs. volume curve.

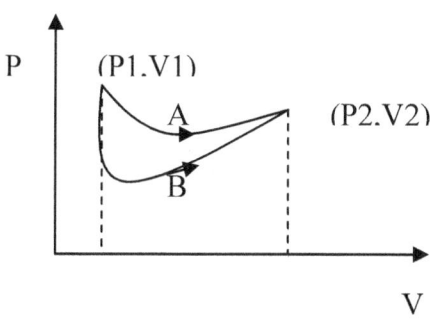

The work done by a gas is not determined only by its initial and final states but by the path through which the gas expands in the PV space. In the example diagram displayed above, a gas starts at pressure P1 and volume V1 and expands to volume V2 at pressure P2. It is clear that if the gas expands through the path A, the area under the curve and the work done by the gas will be greater than the case where the gas expands through the path B.

Problem: If a gas expands from 1 liter to 2 liters at a constant pressure of 3 atmospheres, what is the work done by the gas?

Solution: Since pressure is constant, $W = P \int_{V1}^{V2} dV = P(V2 - V1) = 3(2-1) = 3$ L.atm.

Given that $1L = 10^{-3} m^3$ and 1 atm = $101.3 \times 10^3 N/m^2$, $W = 303.9 J$.

Skill 9   **Second law of thermodynamics (entropy and disorder, reversible and irreversible processes, spontaneity, heat engines, Carnot cycle, efficiency)**

To understand the second law of thermodynamics, we must first understand the concept of entropy. Entropy is the transformation of energy to a more disordered state and is the measure of how much energy or heat is available for work. The greater the entropy of a system, the less energy is available for work. The simplest statement of the second law of thermodynamics is that the entropy of an isolated system not in equilibrium tends to increase over time. The entropy approaches a maximum value at equilibrium. Below are several common examples in which we see the manifestation of the second law.

- The diffusion of molecules of perfume out of an open bottle
- Even the most carefully designed engine releases some heat and cannot convert all the chemical energy in the fuel into mechanical energy
- A block sliding on a rough surface slows down
- An ice cube sitting on a hot sidewalk melts into a little puddle; we must provide energy to a freezer to facilitate the creation of ice

When discussing the second law, scientists often refer to the "arrow of time". This is to help us conceptualize how the second law forces events to proceed in a certain direction. To understand the direction of the arrow of time, consider some of the examples above; we would never think of them as proceeding in reverse. That is, as time progresses, we would never see a puddle in the hot sun spontaneously freeze into an ice cube or the molecules of perfume dispersed in a room spontaneously re-concentrate themselves in the bottle. The above-mentioned examples are **spontaneous** as well as **irreversible**, both characteristic of increased entropy. Entropy change is zero for a **reversible process**, a process where infinitesimal quasi-static changes in the absence of dissipative forces can bring a system back to its original state without a net change to the system or its surroundings. All real processes are irreversible. The idea of a reversible process, however, is a useful abstraction that can be a good approximation in some cases.

The second law of thermodynamics may also be stated in the following ways:
1. No machine is 100% efficient.
2. Heat cannot spontaneously pass from a colder to a hotter object.

If we consider a **heat engine** that absorbs heat Q_h from a hot reservoir at temperature T_h and does work W while rejecting heat Q_c to a cold reservoir at a lower temperature T_c, $Q_h - Q_c = W$. The efficiency of the engine is the ratio of the work done to the heat absorbed and is given by

$$\varepsilon = \frac{W}{Q_h} = \frac{Q_h - Q_c}{Q_h} = 1 - \frac{Q_c}{Q_h}$$

It is impossible to build a heat engine with 100% efficiency, i.e. one where $Q_c = 0$.

Carnot described an ideal reversible engine, the **Carnot engine**, that works between two heat reservoirs in a cycle known as the **Carnot cycle** which consists of two isothermal (12 and 34) and two adiabatic processes (23 and 41) as shown in the diagram below.

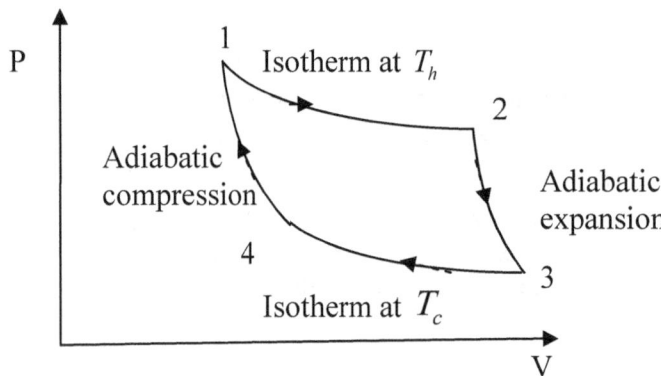

The efficiency of a Carnot engine is given by $\varepsilon = 1 - \dfrac{T_c}{T_h}$ where the temperature values are absolute temperatures. This is the highest efficiency that any engine working between T_c and T_h can reach.

According to **Carnot's theorem**, no engine working between two heat reservoirs can be more efficient than a reversible engine. All such reversible engines have the same efficiency.

Skill 10   Third law and zeroth law of thermodynamics (absolute zero of temperature, law of equilibrium)

The third law of thermodynamics deals with the impossibility of reaching a temperature of absolute zero and a perfectly ordered system. One common way of stating the third law of thermodynamics is, "The entropy of a perfect crystal of an element at the absolute zero of temperature is zero." The law states that, as a system approaches absolute zero, the system will become more orderly. Derivative from this law is the fact that the entropy of a given system near absolute zero is dependent only upon the temperature. At a temperature of absolute zero, there is no thermal energy or heat, so there is no molecular motion. When none of the atoms of a perfectly ordered crystal moves at all, there can be no disorder, no different possible states, and no entropy. The law also provides a reference point for all other entropy values of the system. The standard value of S^0, the entropy of a substance, is actually the integral of entropy from zero to the standard temperature 298.15 K. The values of standard entropies of elements and compounds at standard temperature are all based upon the third law of thermodynamics.

The zeroth law of thermodynamics generally deals with bodies in thermal equilibrium with each other and is the basis for the idea of temperature. Most commonly, the law is stated as, "If two thermodynamic systems are in thermal equilibrium with a third, they are also in thermal equilibrium with each other." Said another way, this very basic law simply states that if object A is same temperature as object B, and object C is the same temperature as object B, then object A and C are also the same temperature. Thermal equilibrium can thus be described as being transitive. Thermal equilibrium exists between two objects when neither object is changing temperature, or when the temperature of the objects is identical. The objects need not be in thermal contact with one another if one is certain that contact would not change their thermal properties.

Skill 11 Energy and energy transformations (kinetic, potential, mechanical, sound, magnetic, electrical, light, heat, nuclear, chemical)

Many forms of energy exist all around us. Energy is defined as the ability to do work. If you are able to measure how much work an object does, or how much heat is exchanged, you can determine the amount of energy that is in a system. Energy and work are measured in Joules. As the law of conservation of energy states, energy can be neither created nor destroyed however, it can be transformed from one form to another. Energy cannot be truly "lost" then, although energy may be wasted or not used to perform work.

Some typical forms of energy are mechanical, heat, sound, electrical, light, chemical, nuclear, and magnetic. Energy can be transformed from mechanical to heat by friction. Additionally, mechanical kinetic energy can combine with magnetic energy, to transform into electrical energy, as when a magnet is spun inside a metal coil. Electrical energy is transformed into light and heat energy when light bulb is turned on and the filament begins to glow. A firefly uses phosphorescence to transform chemical energy into light energy. Within mechanical energy, energy can transform between potential and kinetic repeatedly as is the case with a pendulum.

Energy can also be transformed into matter and vice versa. The equation $E=mc^2$ quantifies the relationship between matter and energy. The conversion of mass to other forms of energy can liberate vast amounts of energy, as shown by nuclear reactors and weapons. However, the mass equivalent of a unit of energy is very small, which is why energy loss is not typically measured by weight.

Energy transformations are classified as thermodynamically reversible or irreversible. A reversible process is one in which no energy is dissipated into empty quantum states, or states of energy with increased disorder. The easiest way to explain the concept is to consider a roller coaster car on a track. A reversible energy transformation occurs as the car travels up and down converting potential energy into kinetic and back. Without friction, the transformation is 100% efficient and no energy is wasted, and the transformation is reversible. However, we know that friction generates heat, and that the heat generated cannot be completely recovered as usable energy, which results in the transformation being irreversible.

DOMAIN V. MODERN PHYSICS, ATOMIC AND NUCLEAR STRUCTURE

Skill 1 Nature of the atom (Rutherford scattering, atomic models, Bohr model, atomic spectra)

In the West, the Greek philosophers Democritus and Leucippus first suggested the concept of the atom. They believed that all atoms were made of the same material but that varied sizes and shapes of atoms resulted in the varied properties of different materials. By the 19th century, John Dalton had advanced a theory stating that each element possesses atoms of a unique type. These atoms were also thought to be the smallest pieces of matter which could not be split or destroyed.

Atomic structure began to be better understood when, in 1897, JJ Thompson discovered the electron while working with cathode ray tubes. Thompson realized the negatively charged electrons were subatomic particles and formulated the "**plum pudding model**" of the atom to explain how the atom could still have a neutral charge overall. In this model, the negatively charged electrons were randomly present and free to move within a soup or cloud of positive charge. Thompson likened this to the dried fruit that is distributed within the English dessert plum pudding though the electrons were free to move in his model.

Ernest Rutherford disproved this model with the discovery of the nucleus in 1909. In **Rutherford's alpha scattering** experiments, he found that alpha particles striking a thin gold foil were scattered at large angles which indicated that the positive charge in an atom was concentrated in a small volume. Rutherford proposed a new "**planetary**" **model** of the atom in which electrons orbited around a positively charged nucleus like planets around the sun. Over the next 20 years, protons and neutrons (subnuclear particles) were discovered while additional experiments showed the inadequacy of the planetary model.

As quantum theory was developed and popularized (primarily by Max Planck and Albert Einstein), chemists and physicists began to consider how it might apply to atomic structure. Niels Bohr put forward a model of the atom in which electrons could only orbit the nucleus in circular orbitals with specific distances from the nucleus, energy levels, and angular momentums. In this model, electrons could only make instantaneous "quantum leaps" between the fixed energy levels of the various orbitals. The Bohr model of the atom was altered slightly by Arnold Sommerfeld in 1916 to reflect the fact that the orbitals were elliptical instead of round.

Though the Bohr model is still thought to be largely correct, it was discovered that electrons do not truly occupy neat, cleanly defined orbitals. Rather, they exist as more of an "electron cloud." The work of Louis de Broglie, Erwin Schrödinger, and Werner Heisenberg showed that an electron can actually be located at any distance from the nucleus. However, we can find the *probability* that the electrons exists at given energy levels (i.e., in particular orbitals) and those probabilities will show that the electrons are most frequently organized within the orbitals originally described in the Bohr model.

The quantum structure of the atom describes electrons in discrete energy levels surrounding the nucleus. When an electron moves from a high energy orbital to a lower energy orbital, a quantum of electromagnetic radiation is emitted, and for an electron to move from a low energy to a higher energy level, a quantum of radiation must be absorbed. The particle that carries this electromagnetic force is called a **photon**. The quantum structure of the atom predicts that only photons corresponding to certain wavelengths of light will be emitted or absorbed by atoms. These distinct wavelengths are measured by **atomic spectroscopy**.

In **atomic absorption spectroscopy**, a continuous spectrum (light consisting of all wavelengths) is passed through the element. The frequencies of absorbed photons are then determined as the electrons increase in energy. An **absorption spectrum** in the visible region usually appears as a rainbow of color stretching from red to violet interrupted by a few black lines corresponding to distinct wavelengths of absorption.

In **atomic emission spectroscopy**, the electrons of an element are excited by heating or by an electric discharge. The frequencies of emitted photons are then determined as the electrons release energy. An **emission spectrum** in the visible region typically consists of lines of light at certain colors corresponding to distinct wavelengths of emission. The bands of emitted or absorbed light at these wavelengths are called **spectral lines**. **Each element has a unique line spectrum**. Light from a star (including the sun) may be analyzed to determine what elements are present.

Skill 2   **Atomic and nuclear structure (electrons, protons and neutrons; electron arrangement, isotopes, hydrogen atom energy levels, nuclear forces and binding energy)**

The constituents of an atom include **protons** which have a positive charge, **neutrons** which have no charge, and **electrons** which have a negative charge. Atoms have no net charge and thus have an equal number of protons and electrons. Protons and neutrons are contained in a small volume at the center of the atom called the **nucleus**. Electrons move in the remaining space of the atom and have very little mass—about 1/1800 of the mass of a proton or neutron. Electrons are prevented from flying away from the nucleus by the attraction that exists between opposite electrical charges. This force is known as **electrostatic** or **coulomb** attraction. **Nuclear force**, which holds the nucleus together, is a byproduct of **strong interactions** and is clearly far stronger than electrostatic forces which would otherwise cause the protons in the nucleus to repel one another.

The identity of an **element** depends on the **number of protons** in the nucleus of the atom. This value is called the **atomic number** and it is sometimes written as a subscript before the symbol for the corresponding element. Atoms and ions of a given element that differ in number of neutrons have a different mass and are called **isotopes**. A nucleus with a specified number of protons and neutrons is called a **nuclide**, and a nuclear particle, either a proton or neutron, may be called a **nucleon**. The total number of nucleons is called the **mass number** and may be written as a superscript before the atomic symbol.

$${}^{14}_{6}\text{C}$$ represents an atom of carbon with 6 protons and 8 neutrons.

The **number of neutrons** may be found by **subtracting the atomic number from the mass number**. For example, uranium-235 has 235–92=143 neutrons because it has 235 nucleons and 92 protons. Different isotopes have different natural abundances and have different nuclear properties. Some nuclei are unstable and emit particles and electromagnetic radiation. These emissions from the nucleus are known as **radioactivity**; the unstable isotopes are known as **radioisotopes**; and the nuclear reactions that spontaneously alter them are known as **radioactive decay**. (For a discussion of nuclear reactions and binding energy see section V.7). Beta decay occurs as a consequence of the **weak interaction** force acting within the atomic nucleus.

Quantum #	Radius
$n \to \infty$	$r_\infty \to \infty$
⋮	⋮
$n = 5$	$r_5 = 25a_0$
$n = 4$	$r_4 = 16a_0$
$n = 3$	$r_3 = 9a_0$
$n = 2$	$r_2 = 4a_0$
$n = 1$ ⊕ (H nucleus)	$r_1 = a_0$

An electron may exist at distinct radial distances (r_n) from the nucleus. These distances are proportional to the square of the **principal quantum number**, n. For a hydrogen atom (shown at left), the proportionality constant is called the **Bohr radius** ($a_0 = 5.29 \times 10^{-11}$ m). This value is the mean distance of an electron from the nucleus at the ground state of $n = 1$. The distances of other electron shells are found by the formula:

$$r_n = a_0 n^2.$$

As $n \to \infty$, the electron is no longer part of the hydrogen atom. Ionization occurs and the atom become an H⁺ ion.

A quantum of energy (ΔE) emitted from or absorbed by an electron transition is directly proportional to the frequency of radiation. The proportionality constant between them is **Planck's constant** ($h = 6.63 \times 10^{-34}$ J·s):

$$\Delta E = h\nu \quad \text{and} \quad \Delta E = \frac{hc}{\lambda}.$$

The energy of an electron (E_n) is inversely proportional to its radius from the nucleus. For a hydrogen atom (shown below left), only the principle quantum number determines the energy of an electron by the **Rydberg constant** ($R_H = 2.18 \times 10^{-18}$ J):

$$E_n = -\frac{R_H}{n^2}.$$

Quantum #	Energy
$n \to \infty$	$E_\infty \to 0$
⋮	⋮
$n = 3$	$E_3 = -\dfrac{R_H}{9}$
$n = 2$	$E_2 = -\dfrac{R_H}{4}$
$n = 1$	$E_1 = -R_H$

The Rydberg constant is used to determine the energy of a photon emitted or absorbed by an electron transition from one shell to another in the H atom:

$$\Delta E = R_H \left(\frac{1}{n_{initial}^2} - \frac{1}{n_{final}^2} \right).$$

When a photon is absorbed, n_{final} is greater than $n_{initial}$, resulting in positive values corresponding to an endothermic process. Ionization occurs when sufficient energy is added for the atom to lose its electron from the ground state. This corresponds to an electron transition from $n_{initial} = 1$ to $n_{final} \to \infty$. The Rydberg constant is the energy required to ionize one atom of hydrogen. Photon emission causes negative values corresponding to an exothermic process because $n_{initial}$ is greater than n_{final}.

Planck's constant and the speed of light are often used to express the Rydberg constant in units of s^{-1} or length. The formulas below determine the photon frequency or wavelength corresponding to a given electron transition:

$$v_{photon} = \left(\frac{R_H}{h}\right)\left|\frac{1}{n_{initial}^2} - \frac{1}{n_{final}^2}\right| \quad \text{and} \quad \lambda_{photon} = \frac{1}{\left(\frac{R_H}{hc}\right)\left|\frac{1}{n_{initial}^2} - \frac{1}{n_{final}^2}\right|}.$$

These formulas **relate observed lines in the hydrogen spectrum to individual transitions** from one quantum state to another.

A simple optical spectroscope separates visible light into distinct wavelengths by passing the light through a prism or diffraction grating. When electrons in hydrogen gas are excited inside a discharge tube, the emission spectroscope shown below detects photons at four visible wavelengths.

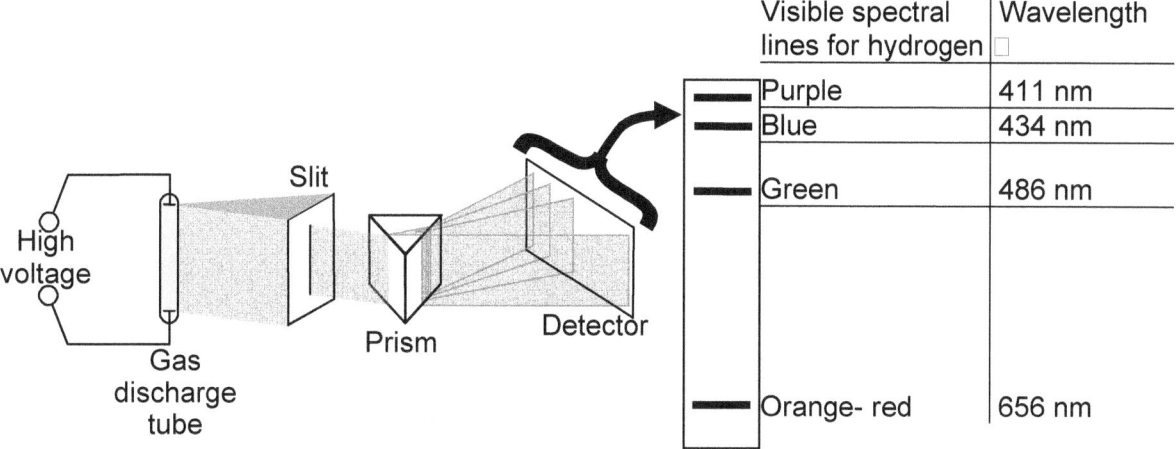

Visible spectral lines for hydrogen	Wavelength
Purple	411 nm
Blue	434 nm
Green	486 nm
Orange- red	656 nm

Every line in the hydrogen spectrum corresponds to a transition between electron energy levels. The four spectral lines from hydrogen emission spectroscopy in the visible range correspond to electron transitions from n = 3, 4, 5, and 6 to n =2 as shown in the table below.

Radiation type	Wavelength $\times$ (nm)	Frequency $\times (s^{-1})$	Energy change $\times E$ (J)	Electron transition $n_{initial} \rightarrow n_{final}$
Ultraviolet	≤ 397	$\geq 7.55 \times 10^{14}$	$\leq -5.00 \times 10^{-19}$	$\infty \rightarrow 1, \ldots 2 \rightarrow 1$ $\infty \rightarrow 2, \ldots 7 \rightarrow 2$
Purple	411	7.31×10^{14}	-4.84×10^{-19}	$6 \rightarrow 2$
Blue	434	6.90×10^{14}	-4.58×10^{-19}	$5 \rightarrow 2$
Green	486	6.17×10^{14}	-4.09×10^{-19}	$4 \rightarrow 2$
Orange-red	656	4.57×10^{14}	-3.03×10^{-19}	$3 \rightarrow 2$
Infrared and beyond	≥ 821	$\leq 3.65 \times 10^{14}$	$\geq -2.42 \times 10^{-19}$	$\infty \rightarrow 3, \ldots 4 \rightarrow 3$ $\infty \rightarrow 4, \ldots 5 \rightarrow 4$ $\vdots$

PHYSICS

Most lines in the hydrogen spectrum are not at visible wavelengths. Larger energy transitions produce ultraviolet radiation and smaller energy transitions produce infrared or longer wavelengths of radiation. Transitions between the first three and the first six energy levels of the hydrogen atom are shown in the diagram to the right. The energy transitions producing the four visible spectral lines are colored grey.

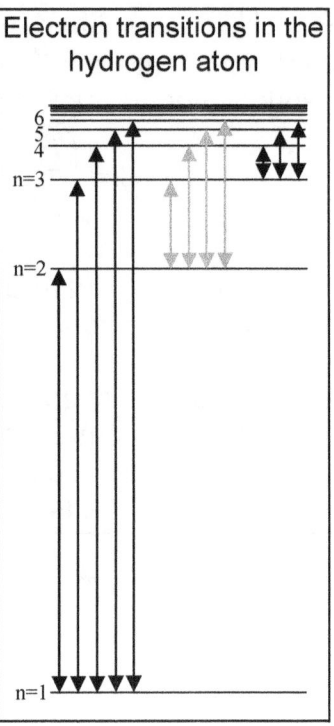
Electron transitions in the hydrogen atom

Skill 3   Radioactivity (radioactive decay, half life, isotopes, decay processes, alpha decay, beta decay, gamma decay, artificial radioactivity)

Some nuclei are unstable and emit particles and electromagnetic radiation. These emissions from the nucleus are known as **radioactivity**. It is often found that some isotopes of an element are radioactive, typically the ones with an excess of neutrons in the nucleus. The unstable isotopes are known as **radioisotopes**; and the nuclear reactions that spontaneously alter them are known as **radioactive decay**. Particles commonly involved in nuclear reactions are listed in the following table:

Particle	Neutron	Proton	Electron	Alpha particle	Beta particle	Gamma rays
Symbol	^1_0n	^1_1p or ^1_1H	$^0_{-1}e$	$^4_2\alpha$ or ^4_2He	$^0_{-1}\beta$ or $^0_{-1}e$	$^0_0\gamma$

Artificial or **induced radioactivity** is the production of radioactive isotopes by bombarding an element with high velocity particles such as neutrons.

In **alpha decay**, an atom emits an alpha particle. An alpha particle contains two protons and two neutrons. This makes it identical to a helium nucleus and so an alpha particle may be written as He^{2+} or it can be denoted using the Greek letter α. Because a nucleus decaying through alpha radiation loses protons and neutrons, the mass of the atom loses about 4 Daltons and it actually becomes a different element (transmutation). For instance:

$$^{238}U \rightarrow \,^{234}Th + \alpha$$

Radioactive heavy nuclei including uranium and radium typically decay by emitting alpha particles. The alpha decay often leaves the nucleus in an excited state with the extra energy subsequently removed by gamma radiation. The energy of alpha particles can be readily absorbed by skin or air and so alpha decaying substances are only harmful to living things if they are introduced internally.

Like alpha decay, **beta decay** involves emission of a particle. In this case, though, it is a beta particle, which is either an electron or positron. Note that a positron is the antimatter equivalent of an electron and so these particles are often denoted β$^-$ and β$^+$. Beta plus and minus decay occur via roughly opposite paths. In beta minus decay, a neutron is converted in a proton (specifically, a down quark is converted to an up quark), an electron and an anti-neutrino; the latter two are emitted. In beta plus decay, on the other hand, a proton is converted to a neutron, a positron, and a neutrino; again, the latter two are emitted. As in alpha decay, a nucleus undergoing beta decay is transmuted into a different element because the number of protons is altered. However, because the total number of nucleons remains unchanged, the atomic mass remains the same (note, that the neutron is actually slightly heavier than a proton so mass is gained during beta plus decay). So examples of beta decay would be:

$$^{137}_{55}Cs \rightarrow \,^{137}_{56}Ba + e^- + \bar{\nu}_e \quad \text{(beta minus decay)}$$

$$^{22}_{11}Na \rightarrow \,^{22}_{10}Ne + e^+ + \nu_e \quad \text{(beta plus decay)}$$

Beta decaying isotopes, such as Strontium 90, are commonly used in cancer treatment. These particles are better able to penetrate skin than alpha particles and so exposure to larger amounts of beta particles poses a risk to all living things.

Gamma radiation is quite different from alpha and beta decay in that it does not involve the emission of nucleon-containing particles or the transmutation of elements. Rather, gamma-ray photons are emitted during gamma decay. These gamma rays are a specific form of electromagnetic radiation that results from certain sub-atomic particle contacts. For instance, electron-positron annihilation leads to the emission of gamma rays. More commonly, though, gamma rays are emitted by nuclei left in an excited state following alpha or beta decay. Thus, gamma decay lowers the energy level of a nucleus, but does not change its atomic mass or charge. The high energy content of gamma rays, coupled with their ability to penetrate dense materials, make them a serious risk to living things.

While the radioactive decay of an individual atom is impossible to predict, a mass of radioactive material will decay at a specific rate. Radioactive isotopes exhibit exponential decay and we can express this decay in a useful equation as follows:

$$A = A_0 e^{kt}$$

Where A is the amount of radioactive material remaining after time t, A_0 is the original amount of radioactive material, t is the elapsed time, and k is the unique activity of the radioactive material. Note that k is unique to each radioactive isotope and it specifies how quickly the material decays. Sometimes it is convenient to express the rate of decay as half-life. **A half-life is the time needed for half a given mass of radioactive material to decay.** Thus, after one half-life, 50% of an original mass will have decayed, after two half lives, 75% will have decayed and so on.

Let's examine a sample problem related to radioactive decay.

Problem: Radiocarbon dating has been used extensively to determine the age of fossilized organic remains. It is based on the fact that while most of the carbon atoms in living things is ^{12}C, a small percentage is ^{14}C. Since ^{14}C is a radioactive isotope, it is lost from a fossilized specimen at a specific rate following the death of an organism. The original and current mass of ^{14}C can be inferred from the relative amount of ^{12}C. So, if the half-life of ^{14}C is 5730 years and a specimen that originally contained 1.28 mg of ^{14}C now contains 0.10 mg, how old is the specimen?

In certain problems, we may be simply provided with the activity, k, but in this problem we must use the information given about half-life to solve for k.

Since we know that after one half-life, 50% of the material remains radioactive, we can plug into the governing equation above:

$$A = A_0 e^{kt}$$

$$0.5\, A_0 = A_0 e^{5730k}$$

$$k = (\ln(0.5))/5730 = -0.0001209$$

Having determined k, we can use this same equation again to determine how old the specimen described above must be:

$$A = A_0 e^{kt}$$

$$0.10 = 1.28 e^{-0.0001209t}$$

$$t = \frac{\ln\left(\frac{0.10}{1.28}\right)}{-0.0001209} = 21087$$

Thus, the specimen is 21,087 years old.

Note that this same equation can be used to calculate the half-life of an isotope if information regarding the decay after a given number of years were provided.

Skill 4   Elementary particles (ionizing radiation)

There are two types of elementary particles: **fermions** and **bosons**. Bosons have integer spin, while fermions have half integer spin.

While there are many subatomic bosons, the most familiar one is the **photon**. A photon has zero mass and charge; in a vacuum, a photon travels at the speed of light. Photons do not spontaneously decay, but can be emitted or absorbed by atoms via a number of natural processes. In fact, photons compose all forms of light and mediate electromagnetic interactions.

The two types of fermions are quarks and leptons.

Quarks: Quarks are found nearly exclusively as components of neutrons and protons. They come in three types of arbitrarily named flavors: up, charm, and top (which have a charge of +2/3) and down, strange, and bottom (which have a charge of –1/3. The flavors have varying mass which must be found via indirect methods.

Leptons: Unlike quarks, leptons do not experience strong nuclear force. There are three flavors of leptons; the muon, the tau, and the electron. Each type of lepton consists of a massive charged particle with the same name as the flavor and a smaller neutral particle. This nearly massless neutral particle is called a **neutrino**. Each lepton has a charge of +1 or –1.

Quarks and leptons combine to form the subatomic particles listed below. In each case, we can compare charge, make-up, mass, location, and mobility of the particle.

Electrons: Electrons are a specific type of lepton that exist outside of the nucleus in positions that are functions of their energy levels. Their arrangement and interactions are key to all chemical bonding and reactions. During such chemical processes, electrons can easily be transferred from one atom to another. The charge of an electron is defined as −1 in atomic units (actual charge is -1.6022×10^{-19} coulomb) and the mass of an electron is $1/1836$ of that of a proton.

Protons: Protons are confined inside the atomic nucleus and have a defined charge of +1 atomic units (1.602×10^{-19} coulomb). The mass of a proton is 1 Dalton (or 1 atomic mass unit). Protons may be lost from the nucleus in certain types of radioactive decay. A proton is composed of 2 up quarks and one down quark.

Neutrons: Like protons, neutrons are confined to the nucleus and only lost during certain types of radioactive decay. Neutrons are uncharged particles with assigned mass of 1 Da (in reality neutrons are slightly heavier than protons). A neutron is composed of 2 down quarks and one up quark.

Finally, for each type of subatomic particle, there is an associated **antiparticle** which has the same mass and opposite charge. Given appropriate quantum conditions, particle/antiparticle pairs can destroy each other.

Ionizing radiation is high energy radiation consisting of particle streams (e.g. products of radioactivity such as electrons or neutrons) or electromagnetic waves (ultraviolet or higher energy) that is able to ionize the atoms and molecules of a medium it is passing through. This kind of radiation is extremely hazardous to human health and damage may be limited by limiting exposure time or using special shields (e.g. x-ray shields used in doctors' and dentists' offices). Ionizing radiation has many industrial applications as well as medical ones such as the use of radiation to treat cancer.

Non- ionizing radiation does not carry enough energy to remove electrons from atoms, and it generally does not cause damage unless it is damage done by heating the material it contacts. Because quantum- events are all- or-nothing, a large amount of non- ionizing radiation is generally harmless, and we can safely live in a world full of lightning flashes and radio waves.

Skill 5   **Organization of matter (elements, compounds, solutions, and mixtures)**

An **element** is a substance that can not be broken down into other substances. To date, scientists have identified 109 elements: 89 are found in nature and 20 are synthetic. An **atom** is the smallest particle of the element that retains the properties of that element. All of the atoms of a particular element are the same. The atoms of each element are different from the atoms of other elements. Elements are assigned an identifying symbol of one or two letters. The symbol for oxygen is O and stands for one atom of oxygen. However, because oxygen atoms in nature are joined together is pairs, the symbol O_2 represents oxygen. This pair of oxygen atoms is a molecule. A **molecule** is the smallest particle of substance that can exist independently and has all of the properties of that substance. A molecule of most elements is made up of one atom. However, oxygen, hydrogen, nitrogen, and chlorine molecules are made of two atoms each.

A **compound** is made of two or more elements that have been chemically combined. Atoms join together when elements are chemically combined. The result is that the elements lose their individual identities when they are joined. The compound that they become has different properties. We use a formula to show the elements of a chemical compound. A **chemical formula** is a shorthand way of showing what is in a compound by using symbols and subscripts. The letter symbols let us know what elements are involved and the number subscript tells how many atoms of each element are involved. No subscript is used if there is only one atom involved. For example, carbon dioxide is made up of one atom of carbon (C) and two atoms of oxygen (O_2), so the formula would be represented as CO_2.

Substances can combine without a chemical change. A **mixture** is any combination of two or more substances in which the substances keep their own properties. A fruit salad is a mixture. So is an ice cream sundae, although you might not recognize each part if it is stirred together. Colognes and perfumes are the other examples. You may not readily recognize the individual elements. However, they can be separated.

Compounds and **mixtures** are similar in that they are made up of two or more substances. However, they have the following opposite characteristics:

   Compounds:
      1. Made up of one kind of particle
      2. Formed during a chemical change
      3. Broken down only by chemical changes
      4. Properties are different from its parts
      5. Has a specific amount of each ingredient.

Mixtures:
1. Made up of two or more particles
2. Not formed by a chemical change
3. Can be separated by physical changes
4. Properties are the same as its parts.
5. Does not have a definite amount of each ingredient.

Common compounds are **acids, bases, salts**, and **oxides** and are classified according to their characteristics.

When two or more pure materials mix in a homogeneous way (with their molecules intermixing on a molecular level), the mixture is called a **solution**. Heterogeneous combinations of materials are called **mixtures**. Dispersions of small particles that are larger than molecules are called **colloids**. Liquid solutions are the most common, but any two phases may form a solution. When a pure liquid and a gas or solid form a liquid solution, the pure liquid is called the **solvent** and the non- liquids are called **solutes**. When all components in the solution were originally liquids, then the one present in the greatest amount is called the solvent and the others are called solutes. Solutions with water as the solvent are called **aqueous** solutions. The amount of solute in a solvent is called its **concentration**. A solution with a small concentration of solute is called **dilute**, and a solution with a large concentration of solute is called **concentrated**.

Skill 6   Physical properties of matter (phase changes, states of matter)

The **phase or state of matter** (solid, liquid, or gas) is identified by its shape and volume. A **solid** has a definite shape and volume. A **liquid** has a definite volume, but no shape. A **gas** has no shape or volume because it will spread out to occupy the entire space of whatever container it is in. While plasma is really a type of gas, its properties are so unique that it is considered a unique phase of matter. **Plasma is a gas that has been ionized**; meaning that at least on electron has been removed from some of its atoms. Plasma shares some characteristics with gas, specifically, the **high kinetic energy** of its molecules. Thus, plasma exists as a diffuse "cloud," though it sometimes includes tiny grains (this is termed dusty plasma). What most distinguishes plasma from gas is that it is **electrically conductive** and exhibits a strong response to electromagnetic fields. This property is a consequence of the **charged particles that result from the removal of electrons** from the molecules in the plasma.

Molecules have **kinetic energy (**they move around), and they also have **intermolecular attractive forces** (they stick to each other). The relationship between these two determines whether a collection of molecules will be a gas, liquid, or solid.

A **gas** has an indefinite shape and an indefinite volume. The kinetic model for a gas is a collection of widely separated molecules, each moving in a random and free fashion, with negligible attractive or repulsive forces between them. Gases will expand to occupy a larger container so there is more space between the molecules. Gases can also be compressed to fit into a small container so the molecules are less separated.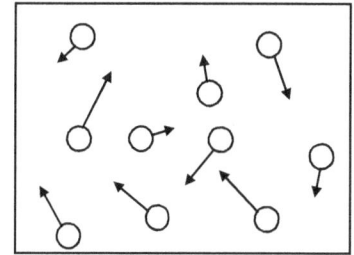
Diffusion occurs when one material spreads into or through another. Gases diffuse rapidly and move from one place to another.

A **liquid** assumes the shape of the portion of any container that it occupies and has a specific volume. The kinetic model for a liquid is a collection of molecules attracted to each other with sufficient strength to keep them close to each other but with insufficient strength to prevent them from moving around randomly. Liquids have a higher density and are much less compressible than gases because the molecules in a liquid are closer together.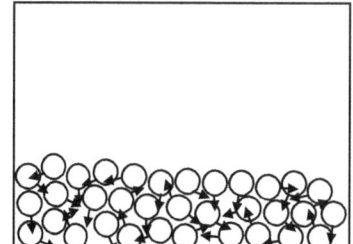

Diffusion occurs more slowly in liquids than in gases because the molecules in a liquid stick to each other and are not completely free to move.

A **solid** has a definite volume and definite shape. The kinetic model for a solid is a collection of molecules attracted to each other with sufficient strength to essentially lock them in place. Each molecule may vibrate, but it has an average position relative to its neighbors. If these positions form an ordered pattern, the solid is called **crystalline**. Otherwise, it is called **amorphous**. Solids have a high density and are almost incompressible because the molecules are close together. Diffusion occurs extremely slowly because the molecules almost never alter their position.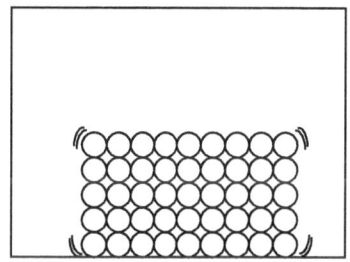

Phase changes occur when the relative importance of kinetic energy and intermolecular forces is altered sufficiently for a substance to change its state. The transition from gas to liquid is called **condensation** and from liquid to gas is called **vaporization**. The transition from liquid to solid is called **freezing** and from solid to liquid is called **melting**. The transition from gas to solid is called **deposition** and from solid to gas is called **sublimation**.

Heat removed from a substance during condensation, freezing, or deposition permits new intermolecular bonds to form, and heat added to a substance during vaporization, melting, or sublimation breaks intermolecular bonds.

During these phase transitions, this **latent heat** is removed or added with **no change in the temperature** of the substance because the heat is not being used to alter the speed of the molecules or the kinetic energy when they strike each other or the container walls. Latent heat alters intermolecular bonds.

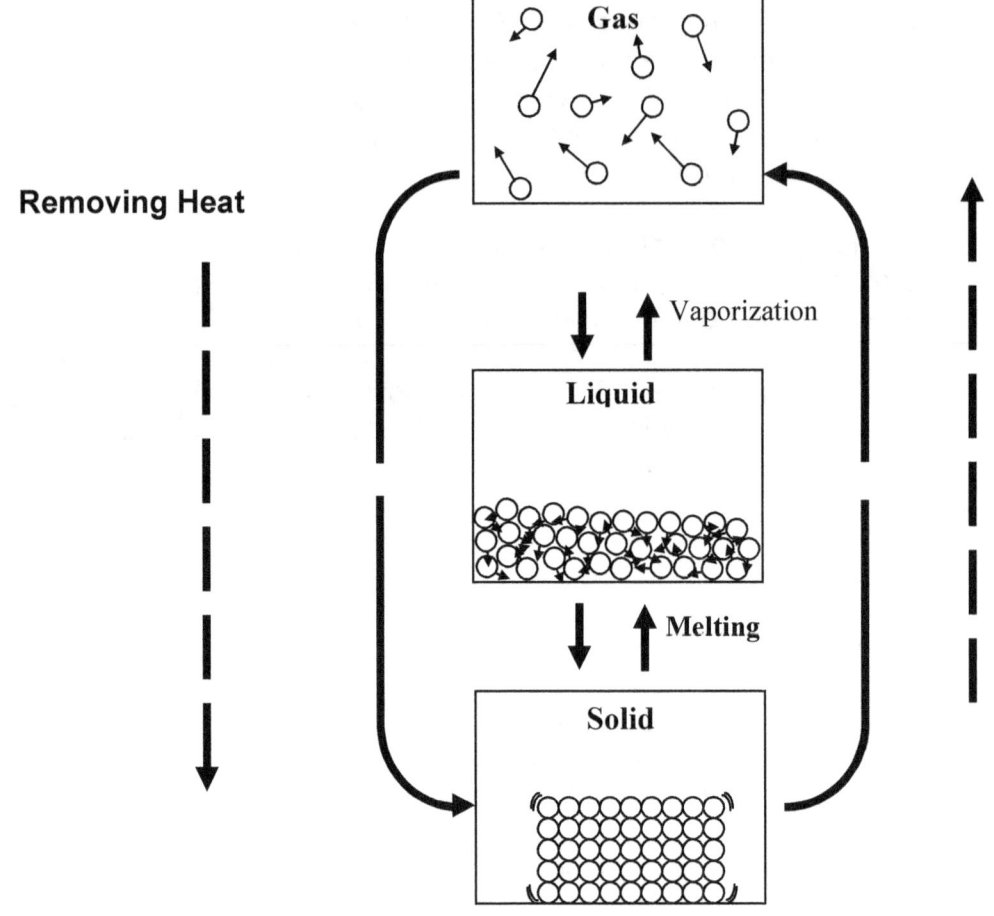

Skill 7   Nuclear energy (fission and fusion, nuclear reactions and their products)

One consequence of the principle of mass-energy equivalence is that mass may be transformed into other types of energy and vice versa. An example of the conversion of mass into energy is the **binding energy** of an atomic nucleus. A deuteron, for instance, has a mass that is less than the sum of the masses its constituent parts, a neutron and a proton. The mass difference is the binding energy that holds the deuteron together.

Problem: The mass of a proton is 1.6726×10^{-27} Kg, the mass of a neutron is 1.6749×10^{-27} Kg, and the mass of a deuteron is 3.3436×10^{-27} Kg. What is the binding energy of a deuteron?

Solution: The difference in mass between the deuteron and its constituents $\Delta m =$

$1.6726 \times 10^{-27} + 1.6749 \times 10^{-27} - 3.3436 \times 10^{-27} = 0.0039 \times 10^{-27}$ Kg

Thus binding energy = $\Delta mc^2 = 3.51 \times 10^{-13} J$

Nuclear **fusion** involves the joining of several nuclei to form one heavier nucleus. Nuclear **fission** is the reverse of fusion, in that it is the splitting of a nucleus to form multiple lighter nuclei. Depending on the weight of the nuclei involved, both fission and fusion may result in either the absorption or release of energy. Iron and nickel have the largest binding energies per nucleon and so are the most stable nuclei. Thus, *fusion* releases energy when the two nuclei are lighter than iron or nickel and *fission* releases energy when the two nuclei are heavier than iron or nickel. Conversely, fusion will absorb energy when the nuclei are heavier and fission will absorb energy when the nuclei are lighter.

Nuclear fusion is common in nature and is the mechanism by which new natural elements are created. Fusion reactions power the stars and (energy absorbing) fusion of heavy elements occurs in supernova explosions. Despite the fact that significant energy is required to trigger the fusion of two nuclei (to overcome the electrostatic repulsion between the positively charged protons), the reaction can be self-sustaining because the energy released by the fusion of two light nuclei is greater than that required to force them together. Fusion is typically much harder to control that fission and so it is not used for power generation though fusion reactions are used to drive hydrogen bombs.

Nuclear fission is unique in that it can be harnessed for a variety of applications. This is done via the use of a chain reaction initiated by the bombardment of certain isotopes with free neutrons. When a nucleus is struck by a free neutron, it splits into smaller nuclei and also produces free neutrons, gamma rays, and alpha and beta particles. The free neutrons can then go onto interact with other nuclei and perpetuate the fission reaction. Isotopes, such as ^{235}U and ^{239}P that sustain the chain reaction are known as fissile and used for nuclear fuel. Because fission can be controlled via chain reaction, it is used in nuclear power generation. Uncontrolled fission reactions are also used in nuclear weapons, including the atomic bombs developed during the Manhattan Project and exploded over Hiroshima and Nagasaki in 1945.

Though it is currently in use in many locations, nuclear fission for power generation remains somewhat controversial. The amount of available energy per pound in nuclear fuel is millions of times that in fossil fuels. Additionally, nuclear power generation does not produce the air and water pollutants that are problematic byproducts of fossil fuel combustion. The currently used fission reactions, however, do produce radioactive waste that must be contained for thousands of years.

Nuclear reactions can consist of simple radioactive decay, fission, fusion, and other nuclear processes. In all cases, both **mass-energy and charge must be conserved**. Below we take a closer look at how this is so in the case of alpha and beta decay.

$$^{238}U \rightarrow {}^{234}Th + \alpha$$

The isotope of uranium in the above reaction weighs 238 Daltons. Because it is uranium, it has 92 protons, meaning it must have 146 neutrons. When it undergoes alpha decay it loses 2 protons and 2 neutrons. The alpha particle weighs 4 Daltons and the nucleus that has undergone decay weighs 234 Daltons. Thus mass is conserved. The decayed nucleus will have a charge reduced by that of 2 protons following the decay. However, the emitted alpha particle carries the additional charge due to its 2 protons. Thus both charge and mass are conserved over all.

$$^{137}_{55}Cs \rightarrow {}^{137}_{56}Ba + e^- + \overline{v}_e$$

This isotope of caesium weighs 137 Daltons and, like all caesium isotopes, it has 55 protons. When it undergoes beta minus decay, a neutron is converted to a proton and an electron and an anti-neutrino are lost. The total mass-energy of the system is conserved since the difference in mass between a neutron and an electron plus proton is balanced by the energy of the emitted electron and the anti-neutrino. In beta minus decay a neutron, with no charge, is split into a positively charged proton and a negatively charged electron. Thus the conservation of charge is satisfied. The electron is emitted, while the proton remains in the nucleus. With one extra proton, the nucleus is now a barium isotope.

$$^{22}_{11}Na \rightarrow {}^{22}_{10}Ne + e^+ + v_e$$

In order to conserve mass-energy of the system, beta plus decay cannot occur in isolation but only in a nucleus since the mass of a neutron is greater than the mass of a proton plus electron. The difference in binding energy of the mother and daughter nucleus provides the additional energy needed for the reaction to go through. Charge is conserved when a positively charged proton is converted into a positively charged positron. With one fewer proton, the decayed nucleus is now a neon isotope weighing 22 Daltons.

Problem: Uranium 235 is used as a nuclear fuel in a chain reaction. The reaction is initiated by a single neutron and produces barium 141, an unknown isotope, and 3 neutrons that can go on to propagate the chain reaction. Determine the unknown isotope. Assume that the kinetic energies and the energy released in the reaction is negligible compared to the masses of the isotopes produced.

Solution: We can begin by writing out the reaction, leaving open the unknown isotope X.

$$^{235}_{92}U + ^{1}_{0}n \rightarrow ^{141}_{56}Ba + X + 3^{1}_{0}n$$

We begin with the charge balance; since the neutron has no charge, the unknown isotope must have 36 protons. Consulting a periodic table, we see that this will mean it is a Krypton isotope. Now we can balance the mass. Since the original nucleus weighed 235 Daltons and one neutron was added to it, the total mass of the resultant nuclei must be 236. So, we can simple subtract the weight of the barium isotope and the 3 new neutrons to find the unknown weight:

236-141-3=92

Thus our unknown isotope is krypton 92, making the balanced equation:

$$^{235}_{92}U + ^{1}_{0}n \rightarrow ^{141}_{56}Ba + ^{92}_{36}Kr + 3^{1}_{0}n$$

Skill 8   Blackbody radiation and photoelectric effect

The wave theory of light explains many different phenomena but falls short when describing effects such as **blackbody radiation** and the **photoelectric effect**.

Blackbody radiation is the characteristic radiation of an ideal blackbody, i.e. a body that absorbs all the radiation incident upon it. Theoretical calculations of the frequency distribution of this radiation using classical physics showed that the energy density of this wave should increase as frequency increases. This result agreed with experiments at shorter wavelengths but failed at large wavelengths where experiment shows that that the energy density of the radiation actually falls back to zero.

In trying to resolve this impasse and derive the spectral distribution of blackbody radiation, Max Planck proposed that an atom can absorb or emit energy only in chunks known as quanta. The energy E contained in each quantum depends on the frequency of the radiation and is given by $E = hf$ where Planck's constant $h = 6.626 \times 10^{-34} J.s = 4.136 \times 10^{-15} eV.s$. Using this quantum hypothesis, Planck was able to provide an explanation for blackbody radiation that matched experiment.

Einstein extended Planck's idea further to suggest that quantization is a fundamental property of electromagnetic radiation which consists of quanta of energy known as **photons**. The energy of each photon is hf where h is Planck's constant.

Problem: A light beam has an intensity of 2W and wavelength of 600nm. What is the energy of each photon in the beam? How many photons are emitted by the beam every second?

Solution: The energy of each photon is given by
$E = hc / \lambda = 6.626 \times 10^{-34} \times 3 \times 10^8 / (600 \times 10^{-9}) = 3.31 \times 10^{-19} J$.
The number of photons emitted each second = $2 / (3.31 \times 10^{-19}) = 6.04 \times 10^{18}$.

Einstein used the photon hypothesis to explain the photoelectric effect.

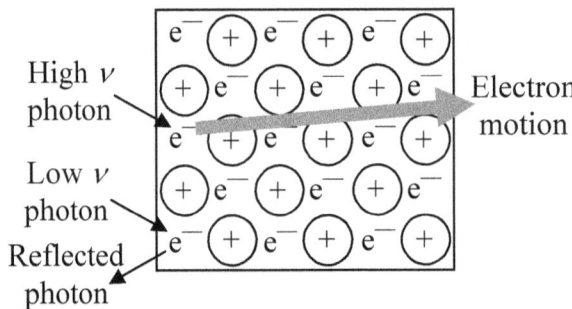

The **photoelectric effect** occurs when **light shining on a clean metal surface causes the surface to emit electrons**. The energy of an absorbed photon is transferred to an electron as shown to the right. If this energy is greater than the binding energy holding the electron close to nearby nuclei then the electron will move. A high energy (high frequency, low wavelength) photon will not only dislodge an electron from the "electron sea" of a metal but it will also impart kinetic energy to the electron, making it move rapidly. These electrons in motion will produce an electric current if a circuit is present.

When the metal surface on which light is incident is a cathode with the anode held at a higher potential V, an electric current flows in the external circuit. It is observed that current flows only for light of higher frequencies. Also there is a threshold negative potential, the **stopping potential** V_0 below which no current will flow in the circuit.

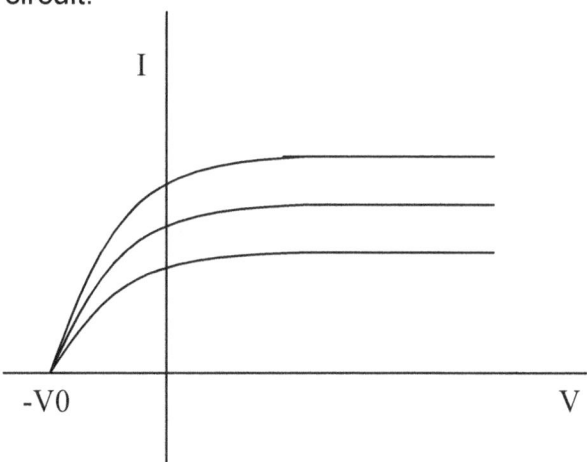

The figure displayed above shows current flow vs. potential for three different intensities of light. It shows that the maximum current flow increases with increasing light intensity but the stopping potential remains the same.

All these observations are counter-intuitive if one considers light to be a wave but may be understood in terms of light particles or photons. According to this interpretation, each photon transfers its energy to a single electron in the metal. Since the energy of a photon depends on its frequency, only a photon of higher frequency can transfer enough energy to an electron to enable it to pass the stopping potential threshold.

When V is negative, only electrons with a kinetic energy greater than $|eV|$ can reach the anode. The maximum kinetic energy of the emitted electrons is given by eV_0. This is expressed by Einstein's photoelectric equation as

$$(\tfrac{1}{2}mv^2)_{max} = eV_0 = hf - \varphi$$

where the **work function** φ is the energy needed to release an electron from the metal and is characteristic of the metal.

Problem: The work function for potassium is 2.20eV. What is the stopping potential for light of wavelength 400nm?

Solution:
$eV_0 = hf - \varphi = hc/\lambda - \varphi = 4.136 \times 10^{-15} \times 3 \times 10^8 / (400 \times 10^{-9}) - 2.20 = 3.10 - 2.20 = 0.90\text{eV}$

Thus stopping potential $V_0 = 0.90\text{V}$

Skill 9   De Broglie's hypothesis and wave-particle duality

The dual wave and particle nature of light has long been considered. In 1924 **Louis de Broglie** suggested that not only light but all matter, particularly electrons, may exhibit wave as well as particle behavior. He proposed that the frequency f and wavelength λ of electron waves are given by the equations

$$f = \frac{E}{h}; \lambda = \frac{h}{p}$$

where p is the momentum of the electron, E is its energy and h is Planck's constant. These are the same relations that Planck proposed for photons. Using deBroglie's equations and considering electrons as standing waves in a circular Bohr orbit, the discrete energy states of an electron could be explained and led to the same set of energy levels found by Bohr. Schrodinger developed these ideas into **wave mechanics**, a general method for finding the quantization condition for a system.

Wave-particle duality is also expressed by **Heisenberg's uncertainty principle** which places a limit on the accuracy with which one can measure the properties of a physical system. This limit is not due to the imperfections of measuring instruments or experimental methods but arises from the fundamental wave-particle duality inherent in quantum systems.

One statement of the uncertainty principle is made in terms of the position and momentum of a particle. If Δx is the uncertainty in the position of a particle in one dimension and Δp the uncertainty in its momentum in that dimension, then according to the uncertainty principle

$$\Delta x \Delta p \geq \hbar / 2$$

where the reduced Planck's constant $\hbar = h/2\pi = 1.05457168 \times 10^{-34} J.s$

Thus if we measure the position of a particle with greater and greater accuracy, at some point the accuracy in the measurement of its momentum will begin to fall. A simple way to understand this is by considering the wave nature of a subatomic particle. If the wave has a single wavelength, then the momentum of the particle is also exactly known using the DeBroglie momentum-wavelength relationship. The position of the wave, however, extends through all space. If waves of several different wavelengths are superposed, the position of the wave becomes increasingly localized as more wavelengths are added. The increased spread in wavelength, however, then results in an increased momentum spread.

An alternate statement of the uncertainty principle may be made in terms of energy and time.

$$\Delta E \Delta t \geq \hbar / 2$$

Thus, for a particle that has a very short lifetime, the uncertainty in the determination of its energy will be large.

Problem: If a proton is confined to a nucleus that is approximately $10^{-15} m$ in diameter, estimate the minimum uncertainty in its momentum in any one dimension.

Solution: The uncertainty of the position of the proton in any dimension cannot be greater than $10^{-15} m$. Using the uncertainty principle we find that the approximate uncertainty in its momentum in any one dimension must be greater than
$\Delta p = \hbar / (2\Delta x) \approx 10^{-19} Kg.m/s$

Skill 10 Special relativity (Michelson-Morley experiment (ether and the speed of light), simultaneity, Lorentz transformations, time dilation, length contraction, velocity addition)

In the nineteenth century it was believed that light propagated through a medium called ether and the speed of light could be measured with respect to this medium. In order to demonstrate the existence of ether, **Michelson and Morley** performed an experiment to determine the relative speed of the earth and the surrounding ether. The set up was a Michelson interferometer in which the speed of light was measured in two different directions, in the direction of the movement of the proposed ether and perpendicular to it. They expected that the difference in the speed of light in the two directions would yield the velocity of the medium. Their experiment, however, had a null result and no detectable difference in speed was found leading to the conclusion that there was probably no ether and that, seemingly, light velocity was independent of the observer's speed.

In 1905 Albert Einstein proposed his **special theory of relativity** that provided an explanation for the results obtained by Michelson and Morley. This formed part of his research article, "On the Electrodynamics of Moving Bodies". Nearly three centuries earlier, Galileo stated that all uniform motion was relative and there was no absolute, well-defined state of rest. Einstein's theory combines Galilean relativity with the idea that all observers will always measure the speed of light to be the same regardless of their linear motion.

This theory has a variety of surprising consequences that seem to violate common sense but were verified experimentally. This theory was called special because it applies the principle of relativity only to inertial frames. Another important thing is that special relativity reveals that c is not just the velocity of a certain phenomenon, light, but rather a fundamental feature of the way space and time are tied together. In particular, special relativity states that it is impossible for any material object to accelerate to light speed.

Postulates: The two postulates of special relativity are
1. Special principle of relativity: The laws of physics are same in all inertial frames of reference which simply means that there are no privileged inertial frames of reference.
2. Invariance of c: The speed of light in a vacuum is a universal constant (c) which is independent of the motion of the light source.

Now let us look at some important consequences of special relativity.
1. Lorentz contraction: Relativity theory depends on "reference frames". An inertial reference frame is a point in space at rest or in uniform motion from which a position can be measured along 3 spatial axes. Lorentz contraction can be described as the perceived reduction of the length L_0 of an object measured by an observer moving relative to the object in the direction of its length. The perceived length is given by

$$L = \sqrt{1 - \frac{v^2}{c^2}} L_0$$

where c is the speed of light and v is the speed of the moving reference frame. This effect is negligible at the speeds we experience everyday but would be noticeable at velocities comparable to that of light.

2. Time dilation: The time lapse between two events is not invariant from one observer to another but is dependent on the relative speeds of the observers' reference frames. Time dilation can also be defined as the difference of time between two ticks in a moving frame and rest frame of the clock. The difference between two ticks measured from a moving frame is larger than the difference between two ticks in the rest frame of a clock. Thus **simultaneity** also becomes a relative concept. Since the time gap between two incidents is not invariant, two events that happen simultaneously in one reference frame may have a time lapse between them in another reference frame. The time lapse observed from a reference frame moving with velocity v relative to the rest frame of a clock is given by

$$\Delta t = \frac{\Delta t_0}{\sqrt{1 - \frac{v^2}{c^2}}}$$

3. Velocity addition: In classical physics, if an object is moving with velocity u' relative to a reference frame that is moving with velocity v (in the same direction), the velocity u of the object with respect to the rest frame is simply a sum of the two velocities v and u'. In a relativistic situation where the velocities are comparable to that of light, however, the postulates of special relativity result in the following relationship between the velocities:

$$u = \frac{u' + v}{1 + vu'/c^2}$$

DOMAIN VI. HISTORY AND NATURE OF SCIENCE; SCIENCE, TECHNOLOGY, AND SOCIAL PERSPECTIVES (STS)

Skill 1 Scientific method of inquiry (formulating problems, formulating and testing hypotheses, making observations, developing generalizations)

The scientific method is a logical set of steps that a scientist goes through to solve a problem. There are as many different scientific methods as there are scientists experimenting. However, there seems to be some pattern to their work. The scientific method is the process by which data is collected, interpreted and validated. While an inquiry may start at any point in this method and may not involve all of the steps here is the general pattern.

Formulating problems

Although many discoveries happen by chance, the standard thought process of a scientist begins with forming a question to research. The more limited and clearly defined the question, the easier it is to set up an experiment to answer it. Scientific questions result from observations of events in nature or events observed in the laboratory. An **observation** is not just a look at what happens. It also includes measurements and careful records of the event. Records could include photos, drawings, or written descriptions. The observations and data collection lead to a question. In physics, observations almost always deal with the behavior of matter. Having arrived at a question, a scientist usually researches the scientific literature to see what is known about the question. Maybe the question has already been answered. The scientist then may want to test the answer found in the literature. Or, maybe the research will lead to a new question.

Sometimes the same observations are made over and over again and are always the same. For example, you can observe that daylight lasts longer in summer than in winter. This observation never varies. Such observations are called **laws** of nature. One of the most important scientific laws was discovered in the late 1700s. Chemists observed that no mass was ever lost or gained in chemical reactions. This law became known as the law of conservation of mass. Explaining this law was a major topic of scientific research in the early 19th century.

Forming a hypothesis

Once the question is formulated, take an educated guess about the answer to the problem or question. This 'best guess' is your hypothesis. A **hypothesis is a statement of a possible answer to the question**. It is a tentative explanation for a set of facts and can be tested by experiments. Although hypotheses are usually based on observations, they may also be based on a sudden idea or intuition.

Experiment

An experiment tests the hypothesis to determine whether it may be a correct answer to the question or a solution to the problem. Some experiments may test the effect of one thing on another under controlled conditions. Such experiments have two variables. The experimenter controls one variable, callred the **independent variable**. The other variable, the **dependent variable**, is the change caused by changing the independent variable. For example, suppose a researcher wanted to test the effect of vitamin A on the ability of rats to see in dim light. The independent variable would be the dose of Vitamin A added to the rats' diet. The dependent variable would be the intensity of light that causes the rats to react. All other factors, such as time, temperature, age, water given to the rats, the other nutrients given to the rats, and similar factors, are held constant. Scientists sometimes do short experiments "just to see what happens". Often, these are not formal experiments. Rather they are ways of making additional observations about the behavior of matter. A good test will try to manipulate as few variables as possible so as to see which variable is responsible for the result. This requires a second example of a **control**. A control is an extra setup in which all the conditions are the same except for the variable being tested.

In most experiments, scientists collect quantitative data, which is data that can be measured with instruments. They also collect qualitative data, descriptive information from observations other than measurements. Interpreting data and analyzing observations are important. If data is not organized in a logical manner, wrong conclusions can be drawn. Also, other scientists may not be able to follow your work or repeat your results.

Conclusion

Finally, a scientist must draw conclusions from the experiment. A conclusion must address the hypothesis on which the experiment was based. The conclusion states whether or not the data supports the hypothesis. If it does not, the conclusion should state what the experiment *did* show. If the hypothesis is not supported, the scientist uses the observations from the experiment to make a new or revised hypothesis. Then, new experiments are planned.

Theory

When a hypothesis survives many experimental tests to determine its validity, the hypothesis may evolve into a **theory**. A theory explains a body of facts and laws that are based on the facts. A theory also reliably predicts the outcome of related events in nature. For example, the law of conservation of matter and many other experimental observations led to a theory proposed early in the 19th century. This theory explained the conservation law by proposing that all matter is made up of atoms which are never created or destroyed in chemical reactions, only rearranged. This atomic theory also successfully predicted the behavior of matter in chemical reactions that had not been studied at the time. As a result, the atomic theory has stood for 200 years with only small modifications.

A theory also serves as a scientific **model**. A model can be a physical model made of wood or plastic, a computer program that simulates events in nature, or simply a mental picture of an idea. A model illustrates a theory and explains nature. For instance, in your science class you may develop a mental (and maybe a physical) model of the atom and its behavior. Outside of science, the word theory is often used to describe someone's unproven notion about something. In science, theory means much more. It is a thoroughly tested explanation of things and events observed in nature. A theory can never be proven true, but it can be proven untrue. All it takes to prove a theory untrue is to show an exception to the theory. The test of the hypothesis may be observations of phenomena or a model may be built to examine its behavior under certain circumstances.

Steps of a Scientific Method

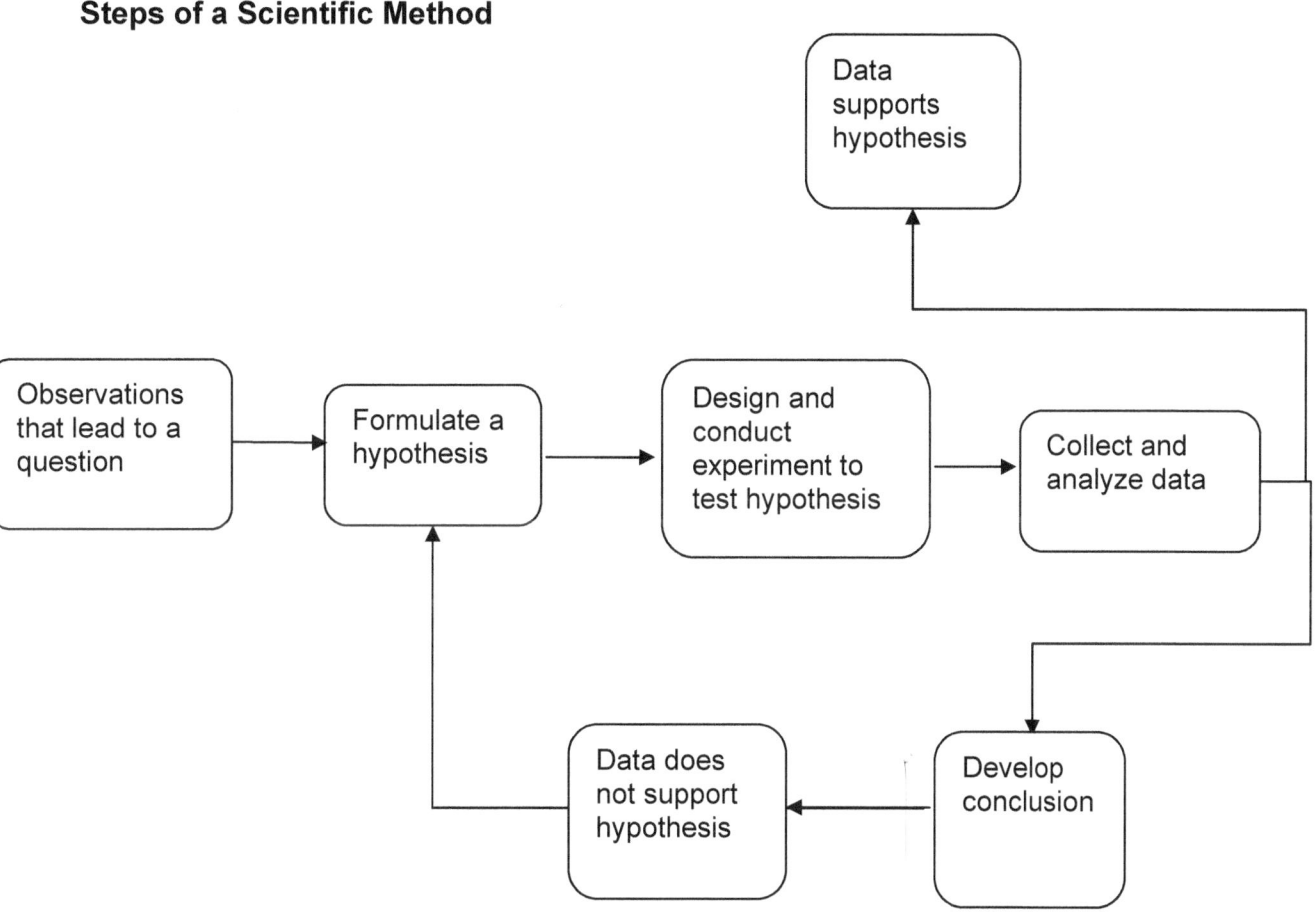

TEACHER CERTIFICATION STUDY GUIDE

Skill 2   Science process skills (observing, hypothesizing, ordering, categorizing, comparing, inferring, applying, communicating)

Application of the scientific method requires familiarity with certain skills that are common to all disciplines. The tools used in each case will depend on the area of study and the specific subject of study. What is common is the mode and attitude with which each skill is applied. Needless to say, uncompromising honesty and reporting of observations with as much objectivity as possible is a fundamental requirement of the scientific process.

Observing: All scientific theories and laws ultimately rest on a strong foundation of experiment. Observation, whether by looking through a microscope or by measuring with a voltmeter, is the fundamental method by which a scientist interacts with the environment to gather the needed data. Scientific observations are not just casual scrutiny but are made in the context of a rigorously planned experiment that specifies precisely what is to be observed and how. Observations must be repeated and the conditions under which they are made clearly noted in order to ensure their validity.

Hypothesizing: Hypothesizing is proposing an answer to a scientific question in order to set the parameters for experiment and to decide what is to be observed and under what conditions. A hypothesis is not a random guess but an educated conjecture based on existing theories and related experiments and a process of rigorous logical reasoning from these basics.

Ordering: For experimental or calculated data to be useful and amenable to analysis, it must be organized appropriately. How data is ordered depends on the question under investigation and the observation process. Ordering may involve prioritizing or categorizing. A data set may be ordered in multiple ways with respect to different variables in order to perform different kinds of analysis on it.

Categorizing: Categorizing is part of the process of ordering or organizing data either into known groups or by identifying new groups through review of the data. The groups may be formed in multiple dimensions, i.e. with respect to more than one variable. For example, a group of objects may be categorized by color as well as size. For some scientific experiments, categorizing may be the goal of the investigation. Categorized data is typically presented in tabular form with the category names as headings.

Comparing: Comparing equivalent quantities is one of the fundamental processes of science. In some cases an observation may be compared with a known or standard number in order to ascertain whether it meets certain criteria. In other cases, data points may be compared with each other for the purpose of prioritizing, categorizing or graphing. Before comparing one must ensure that the numbers are expressed in the same units.

PHYSICS

Inferring: Once data has been organized, graphed and analyzed, a scientist draws conclusions or inferences based on logical reasoning from what he/she sees. An inference is generally drawn in the context of the initial hypothesis. An inference addresses whether the data disproves or supports the hypothesis and to what extent. An inference may also be drawn about some aspect of the data that was not included in the hypothesis. It may lead to the formulation of new problems and new hypotheses or provide answers to questions other than those asked in the hypothesis.

Applying: Applying is the process of connecting a theory, law or thought process to a physical situation, experimental set up or data. Not all laws are applicable to all situations. Also, a theory may be applied to data, for instance, only when it is organized in a specific way. Science requires the ability to evaluate when and how ideas and theories are applicable in specific cases.

Communicating: Communicating, both orally and by writing, is a vital part of scientific activity. Science is never done in a vacuum and theories and experiments are validated only when other scientists can reproduce them and agree with the conclusions. Also, scientific theories can be put to practical use only when others are able to understand them clearly. Thus communication, particularly with peers, is critical to the success of science.

Skill 3 Distinguish among hypotheses, assumptions, models, laws, and theories

<u>Law:</u> A law is a statement of an order or relation of phenomena that, as far as is known, is invariable under the given conditions. Everything we observe in the universe operates according to known natural laws.
- If the truth of a statement is verified repeatedly in a reproducible way then it an reach the level of a natural law.
- Some well known and accepted natural laws of science are:

1. The First Law of Thermodynamics

2. The Second Law of Thermodynamics

3. The Law of Cause and Effect

4. The Law of Biogenesis

5. The Law of Gravity

Theory: In contrast to a law, a scientific theory is used to explain an observation or a set of observations. It is generally accepted to be true, though no real proof exists. The important thing about a scientific theory is that there are no experimental observations to prove it NOT true, and each piece of evidence that exists supports the theory as written. Theories are often accepted at face value since they are often difficult to prove and can be rewritten in order to include the results of all experimental observations. An example of a theory is the big bang theory. While there is no experiment that can directly test whether or not the big bang actually occurred, there is no strong evidence indicating otherwise.

Theories provide a framework to explain the **known** information of the time, but are subject to constant evaluation and updating. There is always the possibility that new evidence will conflict with a current theory.

Some examples of theories that have been rejected because they are now better explained by current knowledge:

Theory of Spontaneous Generation
Inheritance of Acquired Characteristics
The Blending Hypothesis

Some examples of theories that were initially rejected because they fell outside of the accepted knowledge of the time, but are well-accepted today due to increased knowledge and data include:

The sun-centered solar system
Warm-bloodedness in dinosaurs
The germ theory of disease
Continental drift

Assumption: All logical reasoning must start from and build on a set of premises that are accepted without argument and taken to be self-evident. These are known as assumptions and must be explicitly stated before a hypothesis is formulated. Assumptions put boundaries around a scientific question and clarify its parameters.

Hypothesis: A hypothesis is a tentative assumption about the answer to a particular scientific question made in order to draw out and test its logical or empirical consequences. Many refer to a hypothesis as an educated guess about what will happen during an experiment. A hypothesis can be based on prior knowledge and prior observations. It will be proved true or false only through experimentation.

Model: A scientific model is a set of ideas that describes a natural process and is developed by empirical or theoretical methods. Models help scientists focus on the basic fundamental processes. They may be physical representations, such as a space-filling model of a molecule or a map, or they may be mathematical algorithms. Whatever form they take, scientific models are based on what is known about the systems or objects at the time that the models are constructed. Models usually evolve and are improved as scientific advances are made. Sometimes a model must be discarded because new findings show it to be misleading or incorrect.

Models are developed in an effort to explain how things work in nature. Because models are not the "real thing", they can never correctly represent the system or object in all respects. The amount of detail that they contain depends upon how the model will be used as well as the sophistication and skill of the scientist doing the modeling. If a model has too many details left out, its usefulness may be limited. But too many details may make a model too complicated to be useful. So it is easy to see why models lack some features of the real system.

To overcome this difficulty, different models are often used to describe the same system or object. Scientists must then choose which model most closely fits the scientific investigation being carried out, which includes findings that are being described, and, in some cases, which one is compatible with the sophistication of the investigation itself. For example, there are many models of atoms. The solar system model described above is adequate for some purposes because electrons have properties of matter. They have mass and charge and they are found in motion in the space outside the nucleus. However, a highly mathematical model based on the field of quantum mechanics is necessary when describing the energy (or wave) properties of electrons in the atom.

It is on the basis of such models that science makes many of its most important advances because models provide a vehicle for making predictions about the behavior of a system or object. The predictions can then be tested as new measurements, technology or theories are applied to the subject. The new information may result in modification and refinement of the model, although certain issues may remain unresolved by the model for years. The goal, however, is to continue to develop the model in such a way as to move it ever closer to a true description of the natural phenomenon. In this way, models are vital to the scientific process.

Skill 4 **Experimental design (Data collection, interpretation and presentation, significance of controls)**

The design of experiments includes the planning of all steps of the information gathering activity. It is best to start any experiment with a clearly stated and understood set of goals and objectives. This will help lead the experimenter into defining the specific data that needs to be collected, how the data will be collected, and how the data will be analyzed after collection. Specifically, the experimenter should determine the number of observations needed, and over what period. The variables affecting the data collection should also be outlined and a determination of which ones will be held constant, which ones will be varied, and which ones may be out of the experimenter's control.

A scientific control improves the integrity of an experiment by isolating each variable in order to draw conclusions about the effect of the single variable in the result. As much as possible the experiments should be identical, except for the one variable being tested. If there are variables that are beyond the control of the experimenter, it is wise to identify them up front and attempt to mitigate their effect. Controls are generally one of two types, negative and positive. A negative control is used when a negative result is expected in an experiment. The negative control helps correlate a positive result with the variable being tested. For example, testing a drug on a group of rats and maintaining another group who are only given a placebo. A positive control is a sample that is known to produce a positive result to make sure the experiment is working as expected. For example printing a page with he printers own drivers before testing the printer with another program.

Once the data collection process has been defined, collection of information can begin. It is wise to check the data periodically to ensure that the data is reasonable and is being collected appropriately. However, care must be taken that data that does not necessarily fit with expectations is not simply discarded. If the data does not fit expectations, corrections may be needed to the collection method or the expected results may not be accurate.

After collection, experimental data must be analyzed, interpreted and presented in a way that can be understood by others. Data analysis may include statistical methods, curve fitting, and dividing data into subsets. Data analysis transforms the information collected during experimentation with the goal of extracting useful information and drawing conclusions. Data interpretation is the method by which the data, in its raw and analyzed forms is reviewed for meaning and explanation. It is often necessary to look at historical information in the same area of study when interpreting the data from an experiment. Presenting the data is the final step in experimental design. In this way, information is related to others interested or affected by the results of the experiment. Graphical representations are useful to communicate information, although clear and concise language is always necessary to ensure the thorough understanding of your audience.

Skill 5   Integrate the overarching concepts of science

Math, science, and technology all have common themes in how they are applied and understood. Here are some of the fundamental concepts:

Systems
Because the natural world is so complex, the study of science involves the **organization** of items into smaller groups based on interaction or interdependence. These groups are known as **systems**. Systems consist of many separate parts interacting in specific ways to form a whole. It is these interactions that truly define the system. The complete system then has its own characteristics that go beyond the simple collection of its components (i.e., "the whole is more than the sum of the parts"). Natural phenomena and complex technologies can almost always be represented as systems. Examples of systems are the solar system, cardiovascular system, Newton's laws of force and motion, and the laws of conservation.

Models

Science and technology employ models to help simplify concepts. Models can be actual, small, physical mock-ups, mathematical equations, or diagrams that represent the fundamental relationships being studied. Models allow us to gain an understanding of these relationships and to make predictions. Similarly, diagrams, graphs, and charts are often employed to make these phenomena more readily understandable in a visual way.

Change and equilibrium

Another common theme among these three areas is the alternation between change and stability. These alternations occur in natural systems, which typically follow a pattern in which variation is introduced and then equilibrium is restored. Equilibrium is a state in which forces are balanced, resulting in stability. Static equilibrium is stability due to a lack of changes and dynamic equilibrium is stability due to a balance between opposite forces. Similarly, many technologies involve either creation or control of change. The process of change over a long period of time is known as **evolution**. While biological evolution is the most common example, one can also classify technological advancement, changes in the universe, and changes in the environment as evolution.

Scale

In science and technology, we must deal with quantities that have vastly different magnitudes. It is important to understand the relationships between such very different numbers. Specifically, it must be recognized that behavior, and even the laws of physics, may change with scale. Some relationships, for instance, the effect of friction on speed, are only valid over certain size scales. When developing new technology, such as nano-machines, we must keep in mind the importance of scale.

Form and function

Form and function are properties of systems that are closely related. The function of an object usually dictates its form and the form of an object usually facilitates its function. For example, the form of the heart (e.g. muscle, valves) allows it to perform its function of circulating blood through the body. The idea of function dictating the form is also used in architecture.

Skill 6   Historical roots of the physical sciences and the contributions made by major historical figures to the physical sciences

Archimedes

Archimedes was a Greek mathematician, physicist, engineer, astronomer, and philosopher. He is credited with many inventions and discoveries some of which are still in use today such as the Archimedes screw. He designed the compound pulley, a system of pulleys used to lift heavy loads such as ships.

Although Archimedes did not invent the lever, he gave the first rigorous explanation of the principles involved which are the transmission of force through a fulcrum and moving the effort applied through a greater distance than the object to be moved. His Law of the Lever states that magnitudes are in equilibrium at distances reciprocally proportional to their weights.

He also laid down the laws of flotation and described Archimedes' principle which states that a body immersed in a fluid experiences a buoyant force equal to the weight of the displaced fluid.

Amedeo Avogadro

Avogadro was an Italian professor of physics born in the 18th century. He contributed to the understanding of the difference between atoms and molecules and the concept of molarity. The famous Avogadro's principle states that equal volumes of all gases at the same temperature and pressure contain an equal number of molecules.

Niels Bohr

Bohr was a Danish physicist who made fundamental contributions to understanding atomic structure and quantum mechanics. Bohr is widely considered one of the greatest physicists of the twentieth century.

Bohr's model of the atom was the first to place electrons in discrete quantized orbits around the nucleus.

Bohr also helped determine that the chemical properties of an element are largely determined by the number of electrons in the outer orbits of the atom. The idea that an electron could drop from a higher-energy orbit to a lower one emitting a photon of discrete energy originated with Bohr and became the basis for future quantum theory.

He also contributed significantly to the Copenhagen interpretation of quantum mechanics. He received the Nobel Prize for Physics for this work in 1922.

Robert Boyle
Robert Boyle was born in Ireland in 1627 and was one of the most prominent experimentalists of his time. He was the first scientist who kept accurate logs of his experiments and though an alchemist himself, gave birth to the science of chemistry as a separate rigorous discipline. He is well known for Boyle's law that describes the relationship between the pressure and volume of an ideal gas. It was one of the first mathematical expressions of a scientific principle.

Marie Curie
Curie was as a Polish-French physicist and chemist. She was a pioneer in radioactivity and the winner of two Nobel Prizes, one in Physics and the other in Chemistry. She was also the first woman to win the Nobel Prize.

Curie studied radioactive materials, particularly pitchblende, the ore from which uranium was extracted. The ore was more radioactive than the uranium extracted from it which led the Curies (Marie and her husband Pierre) to discover a substance far more radioactive then uranium. Over several years of laboratory work the Curies eventually isolated and identified two new radioactive chemical elements, polonium and radium. Curie refined the radium isolation process and continued intensive study of the nature of radioactivity.

Albert Einstein
Einstein was a German-born theoretical physicist who is widely considered one of the greatest physicists of all time. While best known for the theory of relativity, and specifically mass-energy equivalence, $E = mc^2$, he was awarded the 1921 Nobel Prize in Physics for his explanation of the photoelectric effect and "for his services to Theoretical Physics". In his paper on the photoelectric effect, Einstein extended Planck's hypothesis ($E = h\nu$) of discrete energy elements to his own hypothesis that electromagnetic energy is absorbed or emitted by matter in quanta and proposed a new law $E_{max} = h\nu - P$ to account for the photoelectric effect.

He was known for many scientific investigations including the special theory of relativity which stemmed from an attempt to reconcile the laws of mechanics with the laws of the electromagnetic field. His general theory of relativity considered all observers to be equivalent, not only those moving at a uniform speed. In general relativity, gravity is no longer a force, as it is in Newton's law of gravity, but is a consequence of the curvature of space-time.

Other areas of physics in which Einstein made significant contributions, achievements or breakthroughs include relativistic cosmology, capillary action, critical opalescence, classical problems of statistical mechanics and problems in which they were merged with quantum theory (leading to an explanation of the Brownian movement of molecules), atomic transition probabilities, the quantum theory of a monatomic gas, the concept of the photon, the theory of radiation (including stimulated emission), and the geometrization of physics.

Einstein's research efforts after developing the theory of general relativity consisted primarily of attempts to generalize his theory of gravitation in order to unify and simplify the fundamental laws of physics, particularly gravitation and electromagnetism, which he referred to as the Unified Field Theory.

Michael Faraday
Faraday was an English chemist and physicist who contributed significantly to the fields of electromagnetism and electrochemistry. He established that magnetism could affect rays of light and that the two phenomena were linked. It was largely due to his efforts that electricity became viable for use in technology. The unit for capacitance, the farad, is named after him as is the Faraday constant, the charge on a mole of electrons (about 96,485 coulombs). Faraday's law of induction states that a magnetic field changing in time creates a proportional electromotive force.

Sir Isaac Newton
Newton was an English physicist, mathematician, astronomer, alchemist, and natural philosopher in the late 17th and early 18th centuries. He described universal gravitation and the three laws of motion laying the groundwork for classical mechanics. He was the first to show that the motion of objects on earth and in space is governed by the same set of mechanical laws. These laws became central to the scientific revolution that took place during this period of history. Newton's three laws of motion are:

> I. Every object in a state of uniform motion tends to remain in that state of motion unless an external force is applied to it.
> II. The relationship between an object's mass m, its acceleration a, and the applied force F is $F = ma$.
> III. For every action there is an equal and opposite reaction.

In mechanics, Newton developed the basic principles of conservation of momentum. In optics, he invented the reflecting telescope and discovered that the spectrum of colors seen when white light passes through a prism is inherent in the white light and not added by the prism as previous scientists had claimed. Newton notably argued that light is composed of particles. He also formulated an experimental law of cooling, studied the speed of sound, and proposed a theory of the origin of stars.

J. Robert Oppenheimer
Oppenheimer was an American physicist, best known for his role as the scientific director of the Manhattan Project, the effort to develop the first nuclear weapons. Sometimes called "the father of the atomic bomb", Oppenheimer later lamented the use of atomic weapons. He became a chief advisor to the United States Atomic Energy Commission and lobbied for international control of atomic energy. Oppenheimer was one of the founders of the American school of theoretical physics at the University of California, Berkeley. He did important research in theoretical astrophysics, nuclear physics, spectroscopy, and quantum field theory.

Wilhelm Ostwald
Wilhelm Ostwald, born in 1853 in Latvia, was one of the founders of classical physical chemistry which deals with the properties and reactions of atoms, molecules and ions. He developed the Ostwald process for the synthesis of nitric acid. In 1909 he won the Nobel prize for his work on catalysis, chemical equilibria and reaction velocities.

Linus Pauling
The American chemist Linus Pauling won the Nobel prize for chemistry in 1954 for his investigation of the nature of the chemical bond. He led the way in applying quantum mechanics to chemistry. Later in his career he focused on biochemical problems such as the structure of proteins and sickle cell anemia. He won the Nobel peace prize in 1962 for his contribution to nuclear disarmament.

Skill 7 Scientific knowledge is subject to change

Scientific knowledge is based on a firm foundation of observation. Though mathematics and logic play a major role in defining and deducing scientific theories, ultimately even the most beautiful and intricate theory has to win the support of experiment. A single observation that contradicts an established theory can bring the whole edifice down if confirmed and reproduced. Thus scientific knowledge can never be totally certain and is always open to change based on some new evidence.

Sometimes advanced measuring devices and new equipment make it possible for scientists to detect phenomenon that noone had noted before. Nothing seemed more certain than classical Newtonian physics which explained everything from the motion of the planets to the behavior of earthly objects. At the end of the nineteenth century Lord Kelvin expressed the opinion that physics was complete except for the existence of "two small clouds"; the null result of the Michelson-Morley experiment and the failure of classical physics to predict the spectral distribution of blackbody radiation. The "two small clouds" turned out to be far more significant than Lord Kelvin could have imagined and led to the birth of relativity and quantum theory both of which totally changed the way we see the nature of reality.

If scientific knowledge is not inviolable, what keeps it from being vulnerable to challenge from anybody who thinks they have evidence to contradict a theory? Even though scientific knowledge is not sacred, the scientific process is. No observation is considered valid unless it can be reproduced by another scientist working independently under the same conditions. The peer-review process ensures that all results reported by a scientist undergo strict scrutiny by others working in the same field. Thus it is the integrity of the scientific process that keeps scientific knowledge, despite its openness to change, firmly grounded in objectivity and logic.

Skill 8   Scientific measurement and notation systems

SI is an abbreviation of the French *Système International d'Unités* or the **International System of Units**. It is the most widely used system of units in the world and is the system used in science. The use of many SI units in the United States is increasing outside of science and technology. There are two types of SI units: **base units** and **derived units**. The base units are:

Quantity	Unit name	Symbol
Length	meter	m
Mass	kilogram	kg
Amount of substance	mole	mol
Time	second	s
Temperature	kelvin	K
Electric current	ampere	A
Luminous intensity	candela	cd

The name "kilogram" occurs for the SI base unit of mass for historical reasons. Derived units are formed from the kilogram, but appropriate decimal prefixes are attached to the word "gram." Derived units measure a quantity that may be **expressed in terms of other units**. Some derived units important for physics are:

Derived quantity	Unit name	Expression in terms of other units	Symbol
Area	square meter	m^2	
Volume	cubic meter	m^3	
	liter	$dm^3 = 10^{-3}\, m^3$	L or l
Mass	unified atomic mass unit	$(6.022 \times 10^{23})^{-1}\, g$	u or Da
Time	minute	60 s	min
	hour	60 min = 3600 s	h
	day	24 h = 86400 s	d
Speed	meter per second	m/s	
Acceleration	meter per second squared	m/s^2	
Temperature*	degree Celsius	K	°C
Mass density	gram per liter	$g/L = 1\, kg/m^3$	
Force	newton	$m \cdot kg/s^2$	N
Pressure	pascal	$N/m^2 = kg/(m \cdot s^2)$	Pa
	standard atmosphere§	101325 Pa	atm
Energy, Work, Heat	joule	$N \cdot m = m^3 \cdot Pa = m^2 \cdot kg/s^2$	J
	nutritional calorie§	4184 J	Cal
Heat (molar)	joule per mole	J/mol	
Heat capacity, entropy	joule per kelvin	J/K	
Heat capacity (molar), entropy (molar)	joule per mole kelvin	J/(mol·K)	
Specific heat	joule per kilogram kelvin	J/(kg·K)	
Power	watt	J/s	W
Electric charge	coulomb	s·A	C
Electric potential, electromotive force	volt	W/A	V
Viscosity	pascal second	Pa·s	
Surface tension	newton per meter	N/m	

*Temperature differences in Kelvin are the same as those differences in degrees Celsius. To obtain degrees Celsius from Kelvin, subtract 273.15. Differentiate *m* and meters (m) by context.

§These are commonly used non-SI units.

Decimal multiples of SI units are formed by attaching a **prefix** directly before the unit and a symbol prefix directly before the unit symbol. SI prefixes range from 10^{-24} to 10^{24}. Common prefixes you are likely to encounter in physics are shown below:

Factor	Prefix	Symbol	Factor	Prefix	Symbol
10^9	giga—	G	10^{-1}	deci—	d
10^6	mega—	M	10^{-2}	centi—	c
10^3	kilo—	k	10^{-3}	milli—	m
10^2	hecto—	h	10^{-6}	micro—	μ
10^1	deca—	da	10^{-9}	nano—	n
			10^{-12}	pico—	p

Example: 0.0000004355 meters is 4.355×10^{-7} m or 435.5×10^{-9} m. This length is also 435.5 nm or 435.5 nanometers.

Example: Find a unit to express the volume of a cubic crystal that is 0.2 mm on each side so that the number before the unit is between 1 and 1000.

Solution: Volume is length X width X height, so this volume is $(0.0002 \text{ m})^3$ or 8×10^{-12} m^3. Conversions of volumes and areas using powers of units of length must take the power into account. Therefore:
$$1 \text{ m}^3 = 10^3 \text{ dm}^3 = 10^6 \text{ cm}^3 = 10^9 \text{ mm}^3 = 10^{18} \text{ μm}^3,$$
The length 0.0002 m is 2×10^2 μm, so the volume is also 8×10^6 μm^3. This volume could also be expressed as 8×10^{-3} mm^3. None of these numbers, however, is between 1 and 1000.

Expressing volume in liters is helpful in cases like these. There is no power on the unit of liters, therefore:
$$1 \text{ L} = 10^3 \text{ mL} = 10^6 \text{ μL} = 10^9 \text{ nL}.$$
Converting cubic meters to liters gives
$$8 \times 10^{-12} \text{ m}^3 \times \frac{10^3 \text{ L}}{1 \text{ m}^3} = 8 \times 10^{-9} \text{ L}.$$ The crystal's volume is 8 nanoliters (8 nL).

Example: Determine the ideal gas constant, R, in L·atm/(mol·K) from its SI value of 8.3144 J/(mol·K).

Solution: One joule is identical to one m^3·Pa (see the table on the previous page).
$$8.3144 \frac{\text{m}^3 \cdot \text{Pa}}{\text{mol} \cdot \text{K}} \times \frac{1000 \text{ L}}{1 \text{ m}^3} \times \frac{1 \text{ atm}}{101325 \text{ Pa}} = 0.082057 \frac{\text{L} \cdot \text{atm}}{\text{mol} \cdot \text{K}}$$

The **order of magnitude** is a familiar concept in scientific estimation and comparison. It refers to a category of scale or size of an amount, where each category contains values of a fixed ratio to the categories before or after. The most common ratio is 10. Orders of magnitude are typically used to make estimations of a number. For example, if two numbers differ by one order of magnitude, one number is 10 times larger than the other. If they differ by two orders of magnitude the difference is 100 times larger or smaller, and so on. It follows that two numbers have the same order of magnitude if they differ by less than 10 times the size.

To estimate the order of manitude of a physical quantity, you round the its value to the nearest power of 10. For example, in estimating the human population of the earth, you may not know if it is 5 billion or 12 billion, but a reasonable order of magnitude estimate is 10 billion. Similarly, you may know that Saturn is much larger than Earth and can estiamte that it has approximately 100 times more mass, or that its mass is 2 orders of magnitude larger. The actual number is 95 times the mass of earth. Below are the dimensions of some familiar objects expressed in orders of magnitude.

Physical Item	Size	Order of Magnitude (meters)
Diameter of a hydrogen atom	100 picometers	10^{-10}
Size of a bacteria	1 micrometer	10^{-6}
Size of a raindrop	1 millimeter	10^{-3}
Width of a human finger	1 centimeter	10^{-2}
Height of Washinton Monument	100 meters	10^{2}
Height of Mount Everest	10 kilometers	10^{4}
Diameter of Earth	10 million meters	10^{7}
One light year	1 light year	10^{16}

Skill 9   **Processes involved in scientific data collection and manipulation (organization of data, significant figures, linear regression)**

Scientists use a variety of tools and technologies to perform tests, collect and display data, and analyze relationships. Data is commonly organized in table format and displayed in graphs. Examples of tools that aid in data gathering, organization and analysis include computer-linked probes, spreadsheets, and graphing calculators.

Scientists use **computer-linked probes** to measure various environmental factors including temperature, dissolved oxygen, pH, ionic concentration, and pressure. The advantage of computer-linked probes, as compared to more traditional observational tools, is that the probes automatically gather data and present it in an accessible format. This property of computer-linked probes eliminates the need for constant human observation and manipulation.

Spreadsheets are often used to organize, analyze, and display data. For example, conservation ecologists use spreadsheets to model population growth and development, apply sampling techniques, and create statistical distributions to analyze relationships. Spreadsheet use simplified data collection and manipulation and allows the presentation of data in a logical and understandable format.

Graphing calculators are another technology with many applications to science. For example, physicists use algebraic functions to analyze the time or space dependence of various processes. Graphing calculators can manipulate algebraic data and create graphs for analysis and observation. In addition, the matrix function of graphing calculators may be used to model certain problems. The use of graphing calculators simplifies the creation of displays such as histograms, scatter plots, and line graphs. Scientists can also transfer data and displays to computers for further analysis. Finally, scientists connect computer-linked probes, used to collect data, to graphing calculators to ease the collection, transmission, and analysis of data.

Linear regression is a common technique used by graphing software to translate tabular data into plots in which continuous lines are displayed even though data is available only at some discrete points. Regression is essentially a filling in of the gaps between data points by making a reasonable estimation of what the in-between values are using a standard mathematical process. In other words, it is finding the "best-fit" curve to represent experimental data. The graphs displayed below show how a straight line or a curve can be fitted to a set of discrete data points. There are many different regression algorithms that may be used and they differ from tool to tool. Some tools allow the user to decide what regression method will be used and what kind of curve (straight line, exponential etc.) will be used to fit the data.

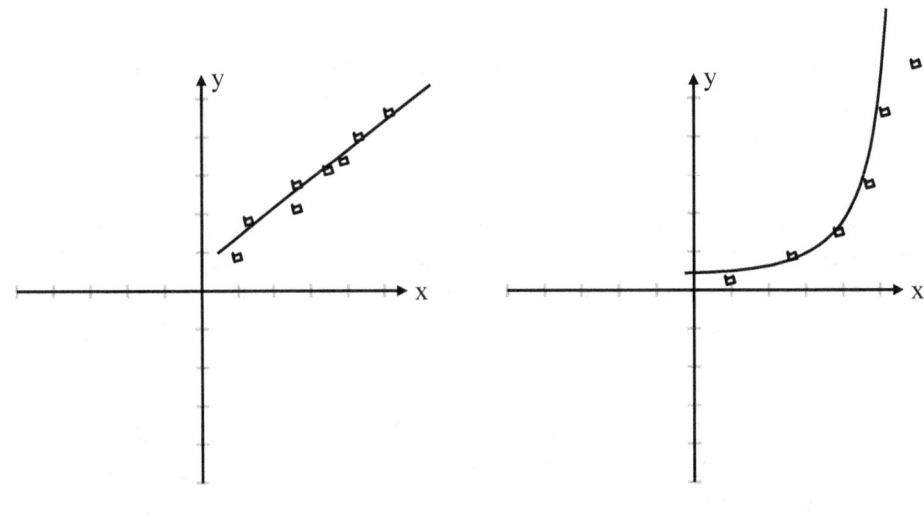

Linear Relationship　　　　　Non- Linear Relationship

Contrast the preceding graphs to the graph of a data set that shows no relationship between variables.

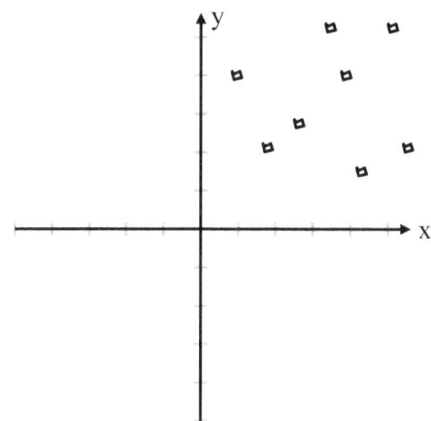

Extrapolation is the process of estimating data points outside a known set of data points. When extrapolating data of a linear relationship, we extend the line of best fit beyond the known values. The extension of the line represents the estimated data points. Extrapolating data is only appropriate if we are relatively certain that the relationship is indeed correctly represented by the best-fit curve.

Significant figures are important concept in the process of measurement and data analysis. They denote the degree of certainty we can have in a given measurement or calculation and reflect the accuracy of the measuring instrument and data-gathering process. The number of significant figures in a quantity is determined by the following rules:

1. Non-zero digits are always significant.
2. Any zeros between two significant digits are significant.
3. A final zero or trailing zeros in the decimal portion <u>ONLY</u> are significant.

For instance, the number of significant figures in the following measurements are as follows:

1) 3.0800   5 significant figures

2) 0.00418   3 significant figures

3) 7.09×10^{-5} 3 significant figures

4) 91,600   3 significant figures

When calculations are performed using experimental data, it is important to record the correct number of significant figures in the result so that information is not lost (by throwing away significant digits) or the accuracy of the measurement is not over-stated (by keeping more digits than are warranted). The following rules apply for operations using data:

1. For multiplication or division, keep the same number of significant figures as the factor with the *fewest* significant figures.

2. For addition or subtraction, keep the same number of decimal places as the term with the *fewest*.

Here are examples of each of these rules:

1) 1.2 x 4.56 = 5.472 but since the first factor has only 2 significant figures, the answer must be also have 2 significant figures and is 5.5
2) 1.234 + 5.67 = 6.904 but since the second term has only 2 decimal places, so must the answer which is 6.90

Skill 10 Interpret and draw conclusions from data, including those presented in tables, graphs, and charts

Trends and patterns in a set of data are most easily identified when the data is displayed graphically. Exact data values in tabular form are also used to calculate various features of the data set. Following are some aspects of data that are commonly analyzed in all scientific disciplines.

The **slope** or the **gradient** of a line is used to describe the rate of change of a variable with respect to another or, in calculus terms, the **derivative** of one variable with respect to another. In the set of examples shown below, the relationship between time, position or distance, velocity and acceleration can be understood conceptually by looking at a graphical representation of each as a function of time. The velocity is the slope of the position vs. time graph and the acceleration is the slope of the velocity vs. time graph.

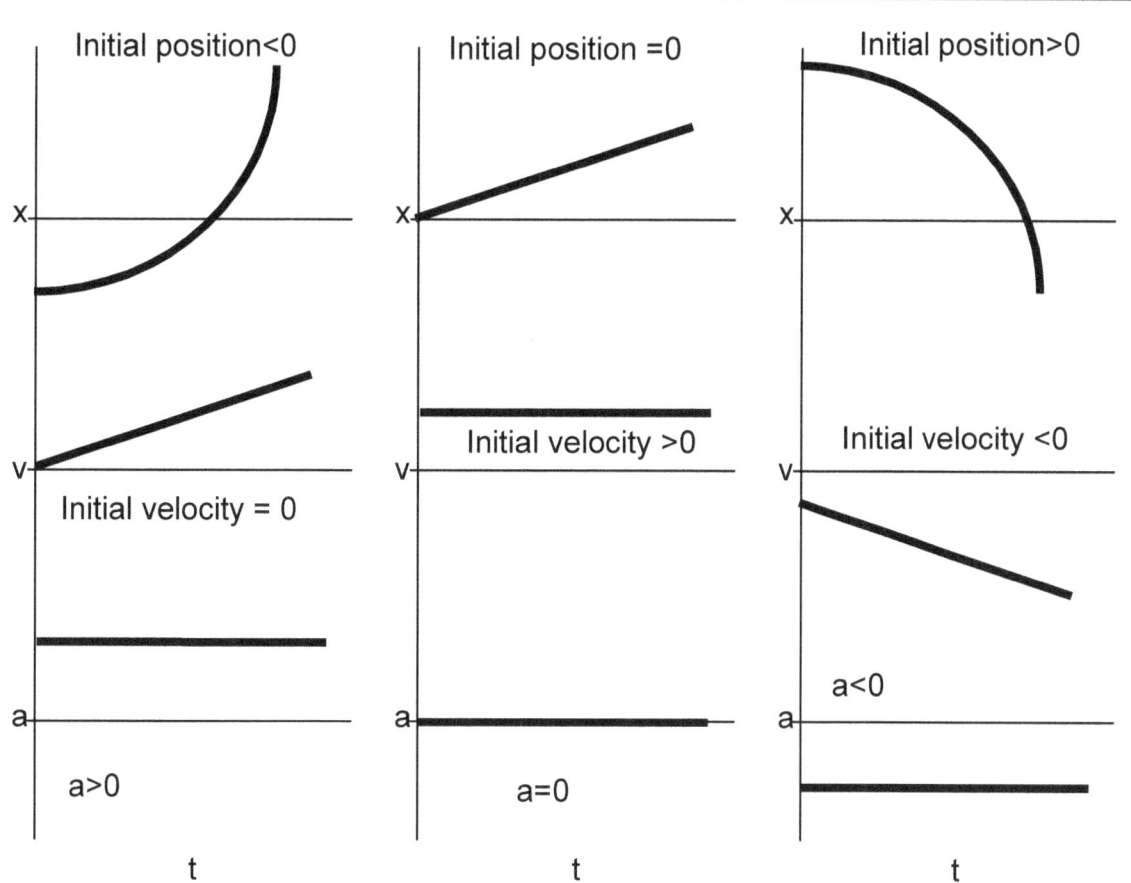

Here are some things we notice by inspecting these graphs:
1) In each case acceleration is constant.
2) A non-zero acceleration produces a position curve that is a parabola.
3) In each case the initial velocity and position are specified separately. The acceleration curve gives the shape of the velocity curve, but not the initial value and the velocity curve gives the shape of the position curve but not the initial position.

The slope of a line may be calculated using the data values displayed in the graph and is given by $m = \dfrac{\Delta y}{\Delta x}$

Given two points (x1, y1) and (x2, y2), the change in x from one to the other is x2 - x1, while the change in y is y2 - y1. Substituting both quantities into the above equation obtains the following:

$$m = \dfrac{y_2 - y_1}{x_2 - x_1}$$

For example: if a line runs through two points: P(1,2) and Q(13,8). By dividing the difference in y-coordinates by the difference in x-coordinates, one can obtain the slope of the line:

$$m = \frac{\Delta y}{\Delta x} = \frac{y_2 - y_1}{x_2 - x_1} = \frac{8 - 2}{13 - 1} = \frac{6}{12} = \frac{1}{2}$$

The slope is 1/2 = 0.5.

The slope of curved lines can be approximated by selecting x and y values that are very close together. In a curved region the slope changes along the curve.

If we let Δx and Δy be the x and y distances between two points on a curve, then Δy /Δx is the slope of a secant line to the curve.

For example, the slope of the secant intersecting
y = x² at (0,0) and (3,9)
is m = (9 - 0) / (3 - 0) = 3

By moving the two points closer together so that Δy and Δx decrease, the secant line more closely approximates a tangent line to the curve, and as such the slope of the secant approaches that of the tangent. In differential calculus, the derivative is essentially taking the change in y with respect to the change in x as the change in x approaches the limit of zero.

The **integral** is usually used to find a measure of totality such as area, volume, mass, or displacement when the rate of change is specified, as in any simple x-y graphical representation. The integral of a function is an extension of the concept of summing and is given by the area under a graphical representation of the function.

The simplest graph to analyze the area under is a flat horizontal line. As an example, let's say f is the constant function f(x) = 3 and we want to find the area under the graph from x= 0 to x=10. This is simply a rectangle 3 units high by 10 units long, or 30 units square. The same result can be found by integrating the function, though this is usually done for more complicated or smooth curves.

Let us imagine the curve of a function f(X) between X=0 and X=10. One way to approximate the area under the curve is to draw numerous rectangles under the curve of a given width, estimate their height and sum the area of each rectangle. We can say that the width of each rectangle is δX. But since the top of each column is not exactly straight, this is only an approximation. When we use integral calculus to determine an integral, we are taking the limit of δX approaching zero, so there will be more and more columns which are thinner and thinner to fill the space between X=0 and X=10. The top of each column then gets closer and closer to being a straight line and our expression for the area therefore gets closer and closer to being exactly right.

Skill 11   **Analyze errors in data that is presented (sources of error, accuracy, precision)**

Scientific data can never be error-free. We can, however, gain useful information from our data by understanding what the sources of error are, how large they are and how they affect our results. Some errors are intrinsic to the measuring instrument, others are operator errors. Errors may be random (in any direction) or systematic (biasing the data in a particular way). In any measurement that is made, data must be quoted along with an estimate of the error in it.

Precision is a measure of how similar repeated measurements from a given device or technique are. Note that this is distinguished from **accuracy** which refers to how close to "correct" a measuring device or technique is. Thus, accuracy can be tested by measuring a known quantity (a standard) and determining how close the value provided by the measuring device is. To determine precision, however, we must make multiple measurements of the same sample. The precision of an instrument is typically given in terms of its standard error or standard deviation. Precision is typically divided into reproducibility and repeatability. These concepts are subtly different and are defined as follows:

Repeatability: Variation observed in measurements made over a short period of time while trying to keep all conditions the same (including using the same instrument, the same environmental conditions, and the same operator)

Reproducibility: Variation observed in measurements taken over a long time period in a variety of different settings (different places and environments, using different instruments and operators)

Both repeatability and reproducibility can be estimated by taking multiple measurements under the conditions specified above. Using the obtained values, standard deviation can be calculated using the formula:

$$\sigma = \sqrt{\frac{1}{N}\sum_{i=1}^{N}(x_i - \overline{x})^2}$$

where σ = standard deviation
        N = the number of measurements
        x_i = the individual measured values
        $\overline{x}$ = the average value of the measured quantity

To obtain a reliable estimate of standard deviation, N, the number of samples, should be fairly large. We can use statistical methods to determine a confidence interval on our measurements. A typical confidence level for scientific investigations is 90% or 95%.

Often in scientific operations we want to determine a quantity that requires many steps to measure. Of course, each time we take a measurement there will be a certain associated error that is a function of the measuring device. Each of these errors contributes to an even greater one in the final value. This phenomenon is known as **propagation of error** or propagation of uncertainty.

A measured value is typically expressed in the form x±Δx, where Δx is the uncertainty or margin of error. What this means is that the value of the measured quantity lies somewhere between x-Δx and x+Δx, but our measurement techniques do not allow us any more precision. If several measurements are required to ultimately decide a value, we must use formulas to determine the total uncertainty that results from all the measurement errors. A few of these formulas for simple functions are listed below:

Formula	Uncertainty
$X = A \pm B$	$(\Delta X)^2 = (\Delta A)^2 + (\Delta B)^2$
$X = cA$	$\Delta X = c \Delta A$
$X = c(A \cdot B)$	$\left(\frac{\Delta X}{X}\right)^2 = \left(\frac{\Delta A}{A}\right)^2 + \left(\frac{\Delta B}{B}\right)^2$
$X = c\left(\frac{A}{B}\right)$	$\left(\frac{\Delta X}{X}\right)^2 = \left(\frac{\Delta A}{A}\right)^2 + \left(\frac{\Delta B}{B}\right)^2$

For example, if we wanted to determine the density of a small piece of metal we would have to measure its weight on a scale and then determine its volume by measuring the amount of water it displaces in a graduated cylinder. There will be error associated with measurements made by both the scale and the graduated cylinder. Let's suppose we took the following measurements:

Mass: 57± 0.5 grams
Volume: 23 ± 3 mm^3

Since density is simply mass divided by the volume, we can determine its value to be:

$$\rho = \frac{m}{V} = \frac{57g}{23mm^3} = 2.5\frac{g}{mm^3}$$

Now we must calculate the uncertainty on this measurement, using the formula above:

$$\left(\frac{\Delta x}{x}\right)^2 = \left(\frac{\Delta A}{A}\right)^2 + \left(\frac{\Delta B}{B}\right)^2$$

$$\Delta x = \left(\sqrt{\left(\frac{\Delta A}{A}\right)^2 + \left(\frac{\Delta B}{B}\right)^2}\right) x = \left(\sqrt{\left(\frac{0.5g}{57g}\right)^2 + \left(\frac{3mm^3}{23mm^3}\right)^2}\right) \times 2.5\frac{g}{mm^3} = 0.3\frac{g}{mm^3}$$

Thus, the final value for the density of this object is 2.5 ± 0.3 g/mm³.

Skill 12 Safety procedures involved in the preparation, storage, use, and disposal of laboratory and field materials

All laboratory solutions should be prepared as directed in the lab manual. Care should be taken to avoid contamination. All glassware should be rinsed thoroughly with distilled water before using and cleaned well after use. All solutions should be made with distilled water as tap water contains dissolved particles that may affect the results of an experiment. Unused solutions should be disposed of according to local disposal procedures.

The "Right to Know Law" covers science teachers who work with potentially hazardous chemicals. Briefly, the law states that employees must be informed of potentially toxic chemicals. An inventory must be made available if requested. The inventory must contain information about the hazards and properties of the chemicals. This inventory is to be checked against the "Substance List". Training must be provided on the safe handling and interpretation of the **Material Safety Data Sheet (MSDS)**.

The following chemicals are potential carcinogens and not allowed in school facilities: Acrylonitriel, Arsenic compounds, Asbestos, Bensidine, Benzene, Cadmium compounds, Chloroform, Chromium compounds, Ethylene oxide, Ortho-toluidine, Nickel powder, and Mercury.

Chemicals should not be stored on bench tops or heat sources. They should be stored in groups based on their reactivity with one another and in protective storage cabinets. All containers within the lab must be labeled. Suspect and known carcinogens must be labeled as such and segregated within trays to contain leaks and spills.

Chemical waste should be disposed of in properly labeled containers. Waste should be separated based on their reactivity with other chemicals.

Biological material should never be stored near food or water used for human consumption. All biological material should be appropriately labeled. All blood and body fluids should be put in a well-contained container with a secure lid to prevent leaking. All biological waste should be disposed of in biological hazardous waste bags.

Material safety data sheets are available for every chemical and biological substance. These are available directly from the company of acquisition or the internet. The manuals for equipment used in the lab should be read and understood before using them.

Skill 13 Identify appropriate use, calibration procedures, and maintenance procedures for laboratory and field equipment

Laboratory and field equipment used for scientific investigation must be handled with the greatest caution and care. The teacher must be completely familiar with the use and maintenance of a piece of equipment before it is introduced to the students. Maintenance procedures for equipment must be scheduled and recorded and each instrument must be calibrated and used strictly in accordance with the specified guidelines in the accompanying manual. Following are some safety precautions one can take in working with different types of equipment:

1. Electricity: Safety in this area starts with locating the main cut off switch. All the power points, switches, and electrical connections must be checked one by one. Batteries and live wires must be checked. All checking must be done with the power turned off. The last act of assembling is to insert the plug and the first act of disassembling is to take off the plug.

2. Motion and forces: All stationary devices must be secured by C-clamps. Protective goggles must be used. Care must be taken at all times while knives, glass rods and heavy weights are used. Viewing a solar eclipse must always be indirect. When using model rockets, NASA's safety code must be implemented.

3. Heat: The master gas valve must be off at all times except while in use. Goggles and insulated gloves are to be used whenever needed. Never use closed containers for heating. Burners and gas connections must be checked periodically. Gas jets must be closed soon after the experiment is over. Fire retardant pads and quality glassware such as Pyrex must be used.

4. Pressure: While using a pressure cooker, never allow pressure to exceed 20 lb/square inch. The pressure cooker must be cooled before it is opened. Care must be taken when using mercury since it is poisonous. A drop of oil on mercury will prevent the mercury vapors from escaping.

5. Light: Broken mirrors or those with jagged edges must be discarded immediately. Sharp-edged mirrors must be taped. Spectroscopic light voltage connections must be checked periodically. Care must be taken while using ultraviolet light sources. Some students may have psychological or physiological reactions to the effects of strobe like (e.g. epilepsy).

6. Lasers: Direct exposure to lasers must not be permitted. The laser target must be made of non-reflecting material. The movement of students must be restricted during experiments with lasers. A number of precautions while using lasers must be taken – use of low power lasers, use of approved laser goggles, maintaining the room's brightness so that the pupils of the eyes remain small. Appropriate beam stops must be set up to terminate the laser beam when needed. Prisms should be set up before class to avoid unexpected reflection.

7. Sound: Fastening of the safety disc while using the high speed siren disc is very important. Teacher must be aware of the fact that sounds higher than 110 decibels will cause damage to hearing.

8. Radiation: Proper shielding must be used while doing experiments with x-rays. All tubes that are used in a physics laboratory such as vacuum tubes, heat effect tubes, magnetic or deflection tubes must be checked and used for demonstrations by the teacher. Cathode rays must be enclosed in a frame and only the teacher should move them from their storage space. Students must watch the demonstration from at least eight feet away.

9. Radioactivity: The teacher must be knowledgeable and properly trained to handle the equipment and to demonstrate. Proper shielding of radioactive material and proper handling of material are absolutely critical. Disposal of any radioactive material must comply with the guidelines of NRC.

TEACHER CERTIFICATION STUDY GUIDE

Skill 14   Preparation of reagents, materials, and apparatus for classroom use

The single most important factor in preparing materials and apparatus for use in the science laboratory is familiarity with the materials and equipment. Teachers must have theoretical as well as hands-on knowledge of how a particular chemical is to be used or how a piece of apparatus functions. Here are some things teachers can do in preparation that will make the laboratory experience safe and effective for students:

- Perform all experiments yourself before introducing them to the students.
- Set up equipment away from table edges and with enough space in between.
- Make sure all safety equipment (e.g. eye wash station) is in place. (See section VI.15 for a list of safety equipment).
- Provide clear written instructions on how to perform an experiment.
- For hazardous chemicals or delicate equipment post warning signs in bold and bright lettering.
- Post multiple copies of laboratory rules that spell out safe ways to do common tasks such as such as handling of bottle stoppers, pouring corrosive reagents, smelling substances.
- Monitor the conditions and stock levels of all materials regularly so that appropriate materials are always on hand.
- When you have a choice in what materials to use, make sure you select the least hazardous ones.

See section VI.12 for information about safety precautions in handling materials and the Material Safety Data Sheets.

Skill 15   Knowledge of safety and emergency procedures for the science classroom and laboratory

Safety is a learned behavior and must be incorporated into instructional plans. Measures of prevention and procedures for dealing with emergencies in hazardous situations have to be in place and readily available for reference. Copies of these must be given to all people concerned, such as administrators and students.

The single most important aspect of safety is planning and anticipating various possibilities and preparing for the eventuality. Any Physics teacher/educator planning on doing an experiment must try it before the students do it. In the event of an emergency, quick action can prevent many disasters. The teacher/educator must be willing to seek help at once without any hesitation because sometimes it may not be clear that the situation is hazardous and potentially dangerous.

There are a number of procedures to prevent and correct any hazardous situation. There are several safety aids available commercially such as posters, safety contracts, safety tests, safety citations, texts on safety in secondary classroom/laboratories, hand books on safety and a host of other equipment. Another important thing is to check the laboratory and classroom for safety and report it to the administrators before staring activities/experiments. It is important that teachers and educators follow these guidelines to protect the students and to avoid most of the hazards. They have a responsibility to protect themselves as well. **There should be not any compromises in issues of safety.**

All science labs should contain the following items of **safety equipment**.
-Fire blanket that is visible and accessible
-Ground Fault Circuit Interrupters (GCFI) within two feet of water supplies
-Signs designating room exits
-Emergency shower providing a continuous flow of water
-Emergency eye wash station that can be activated by the foot or forearm
-Eye protection for every student and a means of sanitizing equipment
-Emergency exhaust fans providing ventilation to the outside of the building
-Master cut-off switches for gas, electric and compressed air. Switches must have permanently attached handles. Cut-off switches must be clearly labeled.
-An ABC fire extinguisher
-Storage cabinets for flammable materials
-Chemical spill control kit
-Fume hood with a motor that is spark proof
-Protective laboratory aprons made of flame retardant material
-Signs that will alert potential hazardous conditions
-Labeled containers for broken glassware, flammables, corrosives, and waste.

Students should wear safety goggles when performing dissections, heating, or while using acids and bases. Hair should always be tied back and objects should never be placed in the mouth. Food should not be consumed while in the laboratory. Hands should always be washed before and after laboratory experiments. In case of an accident, eye washes and showers should be used for eye contamination or a chemical spill that covers the student's body. Small chemical spills should only be contained and cleaned by the teacher. Kitty litter or a chemical spill kit should be used to clean spill. For large spills, the school administration and the local fire department should be notified. Biological spills should only be handled by the teacher. Contamination with biological waste can be cleaned by using bleach when appropriate. Accidents and injuries should always be reported to the school administration and local health facilities. The severity of the accident or injury will determine the course of action to pursue.

Skill 16 Knowledge of the legal responsibilities of the teacher in the science classroom

It is the responsibility of the teacher to provide a safe environment for their students. Proper supervision greatly reduces the risk of injury and a teacher should never leave a class for any reason without providing alternate supervision. After an accident, two factors are considered; **foreseeability** and **negligence**. Foreseeability is the anticipation that an event may occur under certain circumstances. Negligence is the failure to exercise ordinary or reasonable care. Safety procedures should be a part of the science curriculum and a well managed classroom is important to avoid potential lawsuits (See the last few sections for more information about safety). Look up the website of your state Department of Education in order to familiarize yourself with the legal responsibilities of a teacher in your state.

Skill 17 Impact of science and technology on the environment and human affairs

Science and technology have transformed our lives in ways unimagined a generation or two ago. Much of this change has been positive, enriching and prolonging our lives. Technology has increased food production in many parts of the world helping to feed hungry people. New medicines and cures for diseases are unveiled every day. Information technology and the internet have connected people worldwide and empowered individuals while allowing much more communication and cross-pollination of ideas across intellectual disciplines. On the whole, science has made it possible for people to live longer and more productive lives at higher living standards than ever before (of course depending on where they live). The rush towards technological progress, however, has resulted in negative impact, particularly on our natural environment, that no one anticipated or planned for.

The explosion and globalization of technology has resulted in the creation of more industries and increased worldwide travel to such an extent that we are consuming fossil fuels at an unprecedented rate every day. Not only is the dependence on rapidly depleting fossil fuels cause for concern, the emissions created have made global warming a threat to life on the planet. Humans are responsible for the depletion of the ozone layer due to chemicals used for refrigeration and aerosols. Ozone protects the earth from the majority of UV radiation, radiation that promotes skin cancer and has unknown effects on wildlife and plants. Human activity also affects parts of the nutrient cycles by removing nutrients from one part of the biosphere and adding them to another. This results in nutrient depletion in one area and nutrient excess in another. This affects water systems, crops, wildlife, and people. The world's natural water supplies are affected by human use. Oil and wastes from boats and cargo ships pollute the aquatic environment affecting aquatic plant and animal life. Deforestation for urban development has resulted in the extinction or relocation of several species of plants and animals.

There are reasons for optimism in the fact that we have finally begun to recognize how fragile our environment is and how extensively we have been damaging it. More and more scientists and technologists are approaching science from a comprehensive viewpoint and coming up with new environmentally friendly ways to coexist with the natural world without giving up the quest for progress. The Technology Entertainment Design forum (http://www.ted.com/) provides an exciting window into the activities of thought-leaders in all area.

Science and technology are often referred to as a "double- edged sword". Although advances in medicine have greatly improved the quality and length of life, certain moral and ethical controversies have arisen. Advances in science have led to an improved economy through biotechnology as applied to agriculture, yet it has put our health care system at risk and has caused the cost of medical care to skyrocket. Society depends on science and, therefore, it is necessary that the public be scientifically literate and informed. It is important for science teachers to stay abreast of current research and to involve students in critical thinking and ethics whenever possible.

Skill 18   Issues associated with energy production, transmission, management, and use (including nuclear waste removal and transportation)

Energy production and management is an area in which science plays a key role. It is an increasingly important topic in scientific research because of the increasing scarcity of energy yielding resources such as petroleum. With traditional sources of energy becoming more scarce and costly, a major goal of scientific energy research is the creation of alternative, efficient means of energy production. Examples of potential sources of alternative energy include wind, water, solar, nuclear, geothermal, and biomass. An important concern in the production and use of energy, both from traditional and alternative sources, is the effect on the environment and the safe disposal of waste products. Scientific research and study helps determine the best method for energy production, use, and waste product disposal, balancing the need for energy with the associated environmental and health concerns.

Even though there are many potential sources of sustainable and renewable energy, scientists and engineers face multiple practical hurdles in making these available in a large-scale, cost-effective and safe manner. Production costs are often too high and better storage and conversion systems are needed. Transmission and distribution of renewable energy also present challenges of different kinds. Some of the key scientific issues in the area of renewable energy are the need for further investigation of carbon dioxide management, study of photosynthesis, production of cheaper photovoltaic cells, development of systems for storage and use converting solar energy, and more efficient methods for production of ethanol. Energy management also requires foresight in encouraging research into sources that may not be available in the short term but could provide huge benefits in a few decades.

Nuclear energy, though clean and sustainable, is in a class by itself due to the extremely hazardous nature of the production process and waste products. The uncertainties inherent in transporting and storing the waste make it very difficult to predict that impact these activities will have on surrounding communities. Though shipping containers are shielded, spent nuclear fuel emits high levels of gamma and neutron radiation and repeated long-term exposure may present a hazard to handlers, drivers and the general public. The health implications of long exposure to low-level radiation are still not well understood. Most current methods of nuclear waste storage are temporary with no viable long-term options. Long term storage is made difficult by the fact that spent nuclear fuel rods can remain hazardous for thousands of years.

Skill 19 Issues associated with the production, storage, use, management, and disposal of consumer products

An important application of science and technology is the production, storage, use, management, and disposal of consumer products. Scientists from many disciplines work to produce a vast array of consumer products. Genetically modified foods, pharmaceuticals, plastics, nylon, cosmetics, household cleaning products, and color additives are but a few examples of science-based consumer goods.

In addition to consumer product production, science helps determine the proper use and storage of consumer goods. Safe use and storage is a key component of successful production. For example, perishable products like food must be stored and used in a safe and sanitary way. Science helps establish limits and guidelines for the storage and use of perishable food products.

Management and disposal of consumer products is also an important concern. Science helps establish limits for the safe use of potentially hazardous consumer products. For example, household cleaning products are potentially hazardous if used improperly. Scientific testing determines the proper uses and potential hazards of such products.

Finally proper disposal of hazardous waste and recycling of durable materials is important for the health and safety of human populations and the long-term sustainability of the Earth's resources and environment. Until recently the focus in the development of consumer goods has been on satisfying customer needs and desires and ensuring that the goods are safe locally in the short-term. In the coming decades, it will be crucial for scientists to take a long-term and more comprehensive view in creating new materials and products.

Skill 20   Issues associated with the management of natural resources

Nature replenishes itself continually. Natural disturbances such as landslides and brushfires are not only destructive, but, following the destruction, they allow for a new generation of organisms to inhabit the land. For every indigenous organism there exists a natural predator. These predator/prey relationships allow populations to maintain reproductive balance and to not over-utilize food sources, thus keeping food chains in check. Left alone, nature would always find a way to balance itself. Unfortunately, the largest disturbances nature faces are from humans. We have introduced non-indigenous species to many areas, upsetting the predator/prey relationships. Our building has caused landslides and disrupted waterfront ecosystems. We have damaged the ozone layer and over utilized the land entrusted to us. Human activity has had a tremendous impact on the world's natural resources making it imperative that we learn to manage these resources wisely to ensure the future survival of our planet and species. Natural resources are of two types; renewable and non-renewable.

A **renewable resource** is one that is replaced naturally. Living renewable resources are plants and animals. As the population of humans increases resources are used faster. Sometimes renewal of the resource doesn't keep up with the demand. Such is the case with trees. Since the housing industry uses lumber for frames and homebuilding they are often cut down faster than new trees can grow. One way of addressing this problem is through tree farms that use special methods allow trees to grow faster. This is not a complete solution, however, and regulation of deforesting to ensure judicious use of this resource is vital. In the case of animals, cattle are used for their hides and for food. Some animals like deer are killed for sport. Some wild animals are threatened with extinction and need protection on refuges. Each state has an environmental protection agency with divisions of forest management and wildlife management.

Non-living renewable resources include water, air, and soil. Water is renewed in a natural cycle called the water cycle. Oxygen is given off by plants and taken in by animals that in turn expel the carbon dioxide that the plants need. Soil is another renewable resource. Fertile soil is rich in minerals. When plants grow they remove the minerals and make the soil less fertile. Chemical treatments are one way of renewing the soil composition. It is also accomplished naturally when the plants decay back into the soil. The plant material is used to make compost to mix with the soil.

Nonrenewable resources are not easily replaced in a timely fashion. Minerals are nonrenewable resources. Quartz, mica, salt and sulfur are some examples. Mining depletes these resources so society may benefit. Glass is made from from quartz, electronic equipment from mica, and salt has many uses. Sulfur is used in medicine, fertilizers, paper, and matches. Metals are among the most widely used nonrenewable resource. Metals must be separated from the ore. Iron is our most important ore. Gold, silver and copper are often found in a more pure form called native metals.

The most important non-renewable resource that we use up continually in large quantities all over the world is petroleum along with other fossil fuels. These energy sources are so vital to the survival of our economies and societies that a shortage of these will result in a major crisis for human civilization. Thus finding one or more large-scale cost-effective renewable alternative energy sources has become an urgent necessity for our times.

Skill 21   Applications of science and technology in daily life

Science and technology have the center-stage in our daily lives. More and more, it is becoming impossible for people in developed societies to exist without the necessities (e.g. cell phone, home appliances) and conveniences (e.g. satellite TV) afforded by technology. In fact, every day things that used to be conveniences are becoming necessities. Apart from the things that science and technology provide for us, they also represent a mind set and way of thinking such as the application of objectivity or rational thinking in evaluating events and options in our lives. Here are some of the ways in which science and technology are applied in our daily lives:

Health care: In this area, we can see many of the fruits of science and technology in nutrition, genetics, and the development of therapeutic agents. We can see an example of the adaptation of organisms in the development of resistant strains of microbes in response to use of antibiotics. Organic chemistry and biochemistry have been exploited to identify therapeutic targets and to screen and develop new medicines. Advances in molecular biology and our understanding of inheritance have led to the development of genetic screening and allowed us to sequence the human genome.

Environment: There are two broad happenings in environmental science and technology. First, there are many studies being conducted to determine the effects of changing environmental conditions and pollutions. New instruments and monitoring systems have increased the accuracy of these results. Second, advances are being made to mitigate the effects of pollution, develop sustainable methods of agriculture and energy production, and improve waste management.

Agriculture: Development of new technology in agriculture is particularly important as we strive to feed more people with less arable land. Again we see the importance of genetics in developing hybrids that have desirable characteristics. New strains of plants and farming techniques may allow the production of more nutrient rich food and/or allow crops to be grown successfully in harsh conditions. However, it is also important to consider the environmental impact of transgenic species and the use of pesticides and fertilizers. Scientific reasoning and experimentation can assist us in ascertaining the real effect of modern agricultural practices and ways to minimize their impact.

Information technology: The internet has become a new space in our lives. It is the global commons. It is where we conduct business, meet friends, obtain information and find entertainment. It affects all areas of human endeavor by allowing people worldwide to communicate easily and share ideas. It has also spawned a variety of new businesses.

With advances in technology come those in society who oppose it. Ethical questions come into play when discussing issues such as stem cell research or animal research for example. Does it need to be done? What are the effects on humans and animals? The answers to these questions are not always clear and often dependent on circumstance. Is the scientific process of organizing and weighing evidence applicable to these questions or do they lie beyond the domain of science and technology? These are the difficult issues that we have to face as technology moves forward.

Skill 22   Social, political, ethical, and economic issues arising from science and technology

Advances in science and technology create challenges and ethical dilemmas that national governments and society in general must attempt to solve. Local, state, national, and global governments and organizations must increasingly consider policy issues related to science and technology. For example, local and state governments must analyze the impact of proposed development and growth on the environment. Governments and communities must balance the demands of an expanding human population with the local ecology to ensure sustainable growth. Genetic research and manipulation, antibiotic resistance, stem cell research, and cloning are but a few of the issues facing national governments and global organizations today.

In all cases, policy makers must analyze all sides of an issue and attempt to find a solution that protects society while limiting scientific inquiry as little as possible. For example, policy makers must weigh the potential benefits of stem cell research, genetic engineering, and cloning (e.g. medical treatments) against the ethical and scientific concerns surrounding these practices. Many safety concerns have answered by strict government regulations. The FDA, USDA, EPA, and National Institutes of Health are just a few of the government agencies that regulate pharmaceutical, food, and environmental technology advancements

Scientific and technological breakthroughs greatly influence other fields of study and the job market as well. Advances in information technology have made it possible for all academic disciplines to utilize computers and the internet to simplify research and information sharing. In addition, science and technology influence the types of available jobs and the desired work skills. For example, machines and computers continue to replace unskilled laborers and computer and technological literacy is now a requirement for many jobs and careers. Finally, science and technology continue to change the very nature of careers. Because of science and technology's great influence on all areas of the economy, and the continuing scientific and technological breakthroughs, careers are far less stable than in past eras. Workers can thus expect to change jobs and companies much more often than in the past.

Because people often attempt to use scientific evidence in support of political or personal agendas, the ability to evaluate the credibility of scientific claims is a necessary skill in today's society. The media and those with an agenda to advance often overemphasize the certainty and importance of experimental results. One should question any scientific claim that sounds fantastical or overly certain. Scientific, peer-reviewed journals are the most accepted source for information on scientific experiments and studies. Knowledge of experimental design and the scientific method is important in evaluating the credibility of studies. For example, one should look for the inclusion of control groups and the presence of data to support the given conclusions.

Sample Test

DIRECTIONS: Read each item and select the best response.

1. **When acceleration is plotted versus time, the area under the graph represents:**
   (Average Rigor) (Domain I Skill 3)

   A. Time

   B. Distance

   C. Velocity

   D. Acceleration

2. **A skateboarder accelerates down a ramp, with constant acceleration of two meters per second squared, from rest. The distance in meters, covered after four seconds, is:**
   (Rigorous) (Domain I Skill 3)

   A. 10

   B. 16

   C. 23

   D. 37

3. **A brick and hammer fall from a ledge at the same time. They would be expected to:**
   (Easy) (Domain I Skill 4)

   A. Reach the ground at the same time

   B. Accelerate at different rates due to difference in weight

   C. Accelerate at different rates due to difference in potential energy

   D. Accelerate at different rates due to difference in kinetic energy

4. **Gravitational force at the earth's surface causes:**
   (Easy) (Domain I Skill 4)

   A. All objects to fall with equal acceleration, ignoring air resistance

   B. Some objects to fall with constant velocity, ignoring air resistance

   C. A kilogram of feathers to float at a given distance above the earth

   D. Aerodynamic objects to accelerate at an increasing rate

PHYSICS

5. A baseball is thrown with an initial velocity of 30 m/s at an angle of 45°. Neglecting air resistance, how far away will the ball land?
   (Rigorous) (Domain I Skill 4)

   A. 92 m

   B. 78 m

   C. 65 m

   D. 46 m

6. In order to switch between two different reference frames in special relativity, we use the _____ transformation.
   (Average Rigor) (Domain I Skill 5)

   A. Galilean

   B. Lorentz

   C. Euclidean

   D. Laplace

7. A mass is moving at constant speed in a circular path. Choose the true statement below:
   (Average Rigor) (Domain I Skill 6)

   A. Two forces in equilibrium are acting on the mass.

   B. No forces are acting on the mass.

   C. One centripetal force is acting on the mass.

   D. One force tangent to the circle is acting on the mass.

8. The magnitude of a force is:
   (Easy) (Domain I Skill 6)

   A. Directly proportional to mass and inversely to acceleration

   B. Inversely proportional to mass and directly to acceleration

   C. Directly proportional to both mass and acceleration

   D. Inversely proportional to both mass and acceleration

9. If a force of magnitude F gives a mass M an acceleration A, then a force 3F would give a mass 3M an acceleration:
   (Average Rigor) (Domain I Skill 6)

   A. A

   B. 12A

   C. A/2

   D. 6A

10. An inclined plane is tilted by gradually increasing the angle of elevation θ, until the block will slide down at a constant velocity. The coefficient of friction, μ_k, is given by:
    (Rigorous) (Domain I Skill 7)

    A. cos θ

    B. sin θ

    C. cosecant θ

    D. tangent θ

11. A classroom demonstration shows a needle floating in a tray of water. This demonstrates the property of:
    (Easy) (Domain I Skill 8)

    A. Specific Heat

    B. Surface Tension

    C. Oil-Water Interference

    D. Archimedes' Principle

12. A uniform pole weighing 100 grams, that is one meter in length, is supported by a pivot at 40 centimeters from the left end. In order to maintain static position, a 200 gram mass must be placed _____ centimeters from the left end.
    (Rigorous) (Domain I Skill 9)

    A. 10

    B. 45

    C. 35

    D. 50

13. A satellite is in a circular orbit above the earth. Which statement is false?
    (Average Rigor) (Domain I Skill 10)

    A. An external force causes the satellite to maintain orbit.

    B. The satellite's inertia causes it to maintain orbit.

    C. The satellite is accelerating toward the earth.

    D. The satellite's velocity and acceleration are not in the same direction.

14. A 100 g mass revolving around a fixed point, on the end of a 0.5 meter string, circles once every 0.25 seconds. What is the magnitude of the centripetal acceleration?
    (Average Rigor) (Domain I Skill 10)

    A. 1.23 m/s^2

    B. 31.6 m/s^2

    C. 100 m/s^2

    D. 316 m/s^2

15. The kinetic energy of an object is _____ proportional to its _____.
    (Average Rigor) (Domain I Skill 11)

    A. Inversely…inertia

    B. Inversely…velocity

    C. Directly…mass

    D. Directly…time

16. A force is given by the vector 5 N x + 3 N y (where x and y are the unit vectors for the x- and y- axes, respectively). This force is applied to move a 10 kg object 5 m, in the x direction. How much work was done?
    (Rigorous) (Domain I Skill 11)

    A. 250 J

    B. 400 J

    C. 40 J

    D. 25 J

17. An object traveling through air loses part of its energy of motion due to friction. Which statement best describes what has happened to this energy? *(Easy) (Domain I Skill 12)*

   A. The energy is destroyed

   B. The energy is converted to static charge

   C. The energy is radiated as electromagnetic waves

   D. The energy is lost to heating of the air

18. If the internal energy of a system remains constant, how much work is done by the system if 1 kJ of heat energy is added? *(Average Rigor) (Domain I Skill 12)*

   A. 0 kJ

   B. -1 kJ

   C. 1 kJ

   D. 3.14 kJ

19. A mass of 2 kg connected to a spring undergoes simple harmonic motion at a frequency of 3 Hz. What is the spring constant? *(Average Rigor) (Domain I Skill 13)*

   A. 6 kg/s^2

   B. 18 kg/s^2

   C. 710 kg/s^2

   D. 1000 kg/s^2

20. Which statement best describes the relationship of simple harmonic motion to a simple pendulum of length L, mass m and displacement of arc length s?
(Average Rigor) (Domain I Skill 13)

   A. A simple pendulum cannot be modeled using simple harmonic motion

   B. A simple pendulum may be modeled using the same expression as Hooke's law for displacement s, but with a spring constant equal to the tension on the string

   C. A simple pendulum may be modeled using the same expression as Hooke's law for displacement s, but with a spring constant equal to m g/L

   D. A simple pendulum typically does not undergo simple harmonic motion

21. What is the maximum displacement from equilibrium of a 1 kg mass that is attached to a spring with constant k = 100 kg/s^2 if the mass has a velocity of 3 m/s at the equilibrium point?
(Rigorous) (Domain I Skill 13)

   A. 0.1 m

   B. 0.3 m

   C. 3 m

   D. 10 m

22. Which of the following is not an assumption upon which the kinetic-molecular theory of gases is based?
(Rigorous) (Domain I Skill 14)

   A. Quantum mechanical effects may be neglected

   B. The particles of a gas may be treated statistically

   C. The particles of the gas are treated as very small masses

   D. Collisions between gas particles and container walls are inelastic

23. A projectile with a mass of 1.0 kg has a muzzle velocity of 1500.0 m/s when it is fired from a cannon with a mass of 500.0 kg. If the cannon slides on a frictionless track, it will recoil with a velocity of ____ m/s.
    (Rigorous) (Domain I Skill 14)

    A. 2.4

    B. 3.0

    C. 3.5

    D. 1500

24. A car (mass m_1) is driving at velocity v, when it smashes into an unmoving car (mass m_2), locking bumpers. Both cars move together at the same velocity. The common velocity will be given by:
    (Rigorous) (Domain I Skill 14)

    A. m_1v/m_2

    B. m_2v/m_1

    C. $m_1v/(m_1 + m_2)$

    D. $(m_1 + m_2)v/m_1$

25. Which of the following units is not used to measure torque?
    (Average Rigor) (Domain I Skill 15)

    A. slug ft

    B. lb ft

    C. N m

    D. dyne cm

26. Which statement best describes an approach for calculating the kinetic energy of a rotating object?
    (Average Rigor) (Domain I Skill 15)

    A. Rotating objects have no kinetic energy; only objects that undergo linear motion have kinetic energy

    B. Treat the object as a collection of small unit volumes (or particles) and sum the kinetic energies of all the constituent parts

    C. Calculate the rotational inertia, which is equal to the kinetic energy

    D. The kinetic energy of a rotating object cannot be calculated

27. Given the following values for the masses of a proton, a neutron and an alpha particle, what is the nuclear binding energy of an alpha particle?
(Rigorous) (Domain I Skill 16)

Proton mass = 1.6726×10^{-27} kg
Neutron mass = 1.6749×10^{-27} kg
Alpha particle mass = 6.6465×10^{-27} kg

A. 0 J

B. 7.3417×10^{-27} J

C. 4 J

D. 4.3589×10^{-12} J

28. The weight of an object on the earth's surface is designated x. When it is two earth's radii from the surface of the earth, its weight will be:
(Rigorous) (Domain I Skill 17)

A. $x/4$

B. $x/9$

C. $4x$

D. $16x$

29. Given a vase full of water, with holes punched at various heights. The water squirts out of the holes, achieving different distances before hitting the ground. Which of the following accurately describes the situation?
(Average Rigor) (Domain I Skill 20)

A. Water from higher holes goes farther, due to Pascal's Principle.

B. Water from higher holes goes farther, due to Bernoulli's Principle.

C. Water from lower holes goes farther, due to Pascal's Principle.

D. Water from lower holes goes farther, due to Bernoulli's Principle.

30. The electric force in Newtons, on two small objects (each charged to – 10 microCoulombs and separated by 2 meters) is:
(Rigorous) (Domain II Skill 1)

A. 1.0

B. 9.81

C. 31.0

D. 0.225

31. Which of the following is a legitimate explanation for lightning?
    (Average Rigor) (Domain II Skill 2)

    A. Lightning is the result of varying magnetic fields in clouds

    B. Lightning is the result of an electric potential difference greater than the breakdown voltage of air

    C. Lightning is the result of a lens effect due to air masses with different temperatures

    D. Lightning is the result of global warming

32. A hollow conducting sphere of radius R is charged with a total charge Q. What is the magnitude of the electric field at a distance r (given r<R) from the center of the sphere? (k is the electrostatic constant)
    (Rigorous) (Domain II Skill 2)

    A. 0

    B. kQ/R^2

    C. $kQ/(R^2 - r^2)$

    D. $kQ/(R - r)^2$

33. What is the electric flux density (or electric displacement) through each face of a cube, of side length 2 meters, that contains a central point charge of 2 Coulombs?
    (Rigorous) (Domain II Skill 2)

    A. 0.50 Coulomb/m^2

    B. 0.33 Coulomb/m^2

    C. 0.13 Coulomb/m^2

    D. 0.33 Tesla

34. Static electricity generation occurs by:
    (Easy) (Domain II Skill 3)

    A. Telepathy

    B. Friction

    C. Removal of heat

    D. Evaporation

35. A semi-conductor allows current to flow:
    (Easy) (Domain II Skill 3)

    A. Never

    B. Always

    C. As long as it stays below a maximum temperature

    D. When a minimum voltage is applied

36. What should be the behavior of an electroscope, which has been grounded in the presence of a positively charged object (1), after the ground connection is removed and then the charged object is removed from the vicinity (2)?
    (Average Rigor) (Domain II Skill 3)

    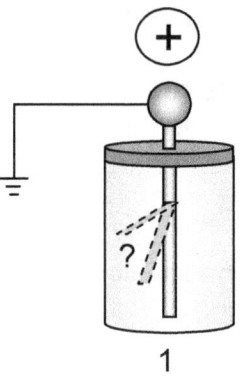 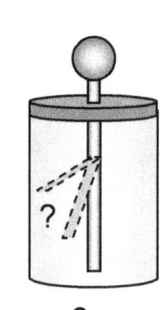

    1   2

    A. The metal leaf will start deflected (1) and then relax to an undeflected position (2)

    B. The metal leaf will start in an undeflected position (1) and then be deflected (2)

    C. The metal leaf will remain undeflected in both cases

    D. The metal leaf will be deflected in both cases

37. All of the following use semi-conductor technology, except a(n):
    (Average Rigor) (Domain II Skill 5)

    A. Transistor

    B. Diode

    C. Capacitor

    D. Operational Amplifier

38. A 10 ohm resistor and a 50 ohm resistor are connected in parallel. If the current in the 10 ohm resistor is 5 amperes, the current (in amperes) running through the 50 ohm resistor is:
    (Rigorous) (Domain II Skill 7)

    A. 1

    B. 50

    C. 25

    D. 60

39. When the current flowing through a fixed resistance is doubled, the amount of heat generated is:
    (Average Rigor) (Domain II Skill 9)

    A. Quadrupled

    B. Doubled

    C. Multiplied by pi

    D. Halved

40. The greatest number of 100 watt lamps that can be connected in parallel with a 120 volt system without blowing a 5 amp fuse is:
    (Rigorous) (Domain II Skill 9)

    A. 24

    B. 12

    C. 6

    D. 1

41. The potential difference across a five Ohm resistor is five Volts. The power used by the resistor, in Watts, is:
    (Rigorous) (Domain II Skill 9)

    A. 1

    B. 5

    C. 10

    D. 20

42. What is the resonant angular frequency (ω) of the following circuit?
    (Average Rigor) (Domain II Skill 10)

    A. 1 Hz

    B. 0.5 Hz

    C. 0.71 radians/sec

    D. This is not a resonant circuit

43. How much power is dissipated through the following resistive circuit? *(Average Rigor) (Domain II Skill 10)*

    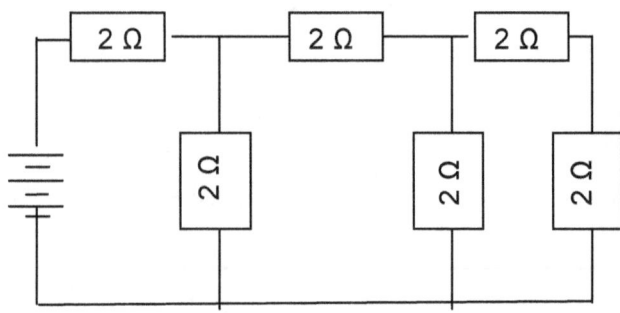

    A. 0 W

    B. 0.22 W

    C. 0.31 W

    D. 0.49 W

44. What effect might an applied external magnetic field have on the magnetic domains of a ferromagnetic material?
    (Rigorous) (Domain II Skill 13)

    A. The domains that are not aligned with the external field increase in size, but those that are aligned decrease in size

    B. The domains that are not aligned with the external field decrease in size, but those that are aligned increase in size

    C. The domains align perpendicular to the external field

    D. There is no effect on the magnetic domains

45. Which of the following statements may be taken as a legitimate inference based upon the Maxwell equation that states $\nabla \cdot \mathbf{B} = 0$?
    (Average Rigor) (Domain II Skill 14)

    A. The electric and magnetic fields are decoupled

    B. The electric and magnetic fields are mediated by the W boson

    C. There are no photons

    D. There are no magnetic monopoles

46. What is the direction of the magnetic field at the center of the loop of current (I) shown below (i.e., at point A)?
    (Easy) (Domain II Skill 15)

    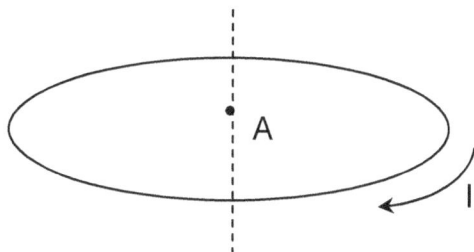

    A. Down, along the axis (dotted line)
    B. Up, along the axis (dotted line)
    C. The magnetic field is oriented in a radial direction
    D. There is no magnetic field at point A

47. What is the effect of running current in the same direction along two parallel wires, as shown below?
(Rigorous) (Domain II Skill 15)

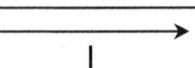

A. There is no effect

B. The wires attract one another

C. The wires repel one another

D. A torque is applied to both wires

48. The current induced in a coil is defined by which of the following laws?
(Easy) (Domain II Skill 16)

A. Lenz's Law

B. Burke's Law

C. The Law of Spontaneous Combustion

D. Snell's Law

49. A static magnetic flux density of 1 Tesla is linked by a wire loop, as shown below. What is the electromotive force (EMF) around the loop?
(Average Rigor) (Domain II Skill 16)

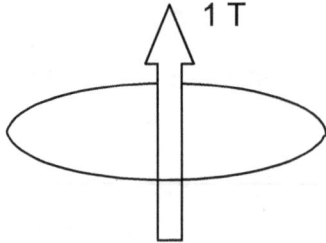

A. 1 Volt

B. 2 Volts

C. 5 Volts

D. 0 Volts

50. A light bulb is connected in series with a rotating coil within a magnetic field. The brightness of the light may be increased by any of the following except:
(Average Rigor) (Domain II Skill 16)

A. Rotating the coil more rapidly.

B. Using more loops in the coil.

C. Using a different color wire for the coil.

D. Using a stronger magnetic field.

51. The use of two circuits next to each other, with a change in current in the primary circuit, demonstrates:
(Rigorous) (Domain II Skill 16)

   A. Mutual current induction

   B. Dielectric constancy

   C. Harmonic resonance

   D. Resistance variation

52. An electromagnetic wave propagates through a vacuum. Independent of its wavelength, it will move with constant:
(Easy) (Domain III Skill 1)

   A. Acceleration

   B. Velocity

   C. Induction

   D. Sound

53. A wave generator is used to create a succession of waves. The rate of wave generation is one every 0.33 seconds. The period of these waves is:
(Average Rigor) (Domain III Skill 1)

   A. 2.0 seconds

   B. 1.0 seconds

   C. 0.33 seconds

   D. 3.0 seconds

54. A wave has speed 60 m/s and wavelength 30,000 m. What is the frequency of the wave?
(Average Rigor) (Domain III Skill 1)

   A. 2.0×10^{-3} Hz

   B. 60 Hz

   C. 5.0×10^2 Hz

   D. 1.8×10^6 Hz

55. **Rainbows are created by:**
(Easy) (Domain III Skill 3)

   A. Reflection, dispersion, and recombination

   B. Reflection, resistance, and expansion

   C. Reflection, compression, and specific heat

   D. Reflection, refraction, and dispersion

56. **The wave phenomenon of polarization applies only to:**
(Average Rigor) (Domain III Skill 4)

   A. Longitudinal waves

   B. Transverse waves

   C. Sound

   D. Light

57. **A vibrating string's frequency is _____ proportional to the _____.**
(Rigorous) (Domain III Skill 4)

   A. Directly; Square root of the tension

   B. Inversely; Length of the string

   C. Inversely; Squared length of the string

   D. Inversely; Force of the plectrum

58. **A stationary sound source produces a wave of frequency F. An observer at position A is moving toward the horn, while an observer at position B is moving away from the horn. Which of the following is true?**
(Rigorous) (Domain III Skill 4)

   A. $F_A < F < F_B$

   B. $F_B < F < F_A$

   C. $F < F_A < F_B$

   D. $F_B < F_A < F$

59. The velocity of sound is greatest in:
    (Average Rigor) (Domain III Skill 5)

    A. Water

    B. Steel

    C. Alcohol

    D. Air

60. The combination of overtones produced by a musical instrument is known as its:
    (Average Rigor) (Domain III Skill 5)

    A. Timbre

    B. Chromaticity

    C. Resonant Frequency

    D. Flatness

61. The following statements about sound waves are true except:
    (Average Rigor) (Domain III Skill 5)

    A. Sound travels faster in liquids than in gases.

    B. Sound waves travel through a vacuum.

    C. Sound travels faster through solids than liquids.

    D. Ultrasound can be reflected by the human body.

62. The highest energy is associated with:
    (Easy) (Domain III Skill 6)

    A. UV radiation

    B. Yellow light

    C. Infrared radiation

    D. Gamma radiation

63. Which of the following apparatus can be used to measure the wavelength of a sound produced by a tuning fork?
    (Average Rigor) (Domain III Skill 7)

    A. A glass cylinder, some water, and iron filings

    B. A glass cylinder, a meter stick, and some water

    C. A metronome and some ice water

    D. A comb and some tissue

64. All of the following phenomena are considered "refractive effects" except for:
    (Rigorous) (Domain III Skill 8)

    A. The red shift

    B. Total internal reflection

    C. Lens dependent image formation

    D. Snell's Law

65. A monochromatic ray of light passes from air to a thick slab of glass (n = 1.41) at an angle of 45° from the normal. At what angle does it leave the air/glass interface?
    (Rigorous) (Domain III Skill 8)

    A. 45°

    B. 30°

    C. 15°

    D. 55°

66. Which of the following is *not* a legitimate explanation for refraction of light rays at boundaries between different media?
    (Rigorous) (Domain III Skill 8)

    A. Light seeks the path of least time between two different points

    B. Due to phase matching and other boundary conditions, plane waves travel in different directions on either side of the boundary, depending on the material parameters

    C. The electric and magnetic fields become decoupled at the boundary

    D. Light rays obey Snell's law

67. If an object is 20 cm from a convex lens whose focal length is 10 cm, the image is:
    (Rigorous) (Domain III Skill 9)

    A. Virtual and upright

    B. Real and inverted

    C. Larger than the object

    D. Smaller than the object

68. Automobile mirrors that have a sign, "objects are closer than they appear" say so because:
    (Rigorous) (Domain III Skill 10)

    A. The real image of an obstacle, through a converging lens, appears farther away than the object.

    B. The real or virtual image of an obstacle, through a converging mirror, appears farther away than the object.

    C. The real image of an obstacle, through a diverging lens, appears farther away than the object.

    D. The virtual image of an obstacle, through a diverging mirror, appears farther away than the object.

69. An object two meters tall is speeding toward a plane mirror at 10 m/s. What happens to the image as it nears the surface of the mirror?
    (Rigorous) (Domain III Skill 10)

    A. It becomes inverted.

    B. The Doppler Effect must be considered.

    C. It remains two meters tall.

    D. It changes from a real image to a virtual image.

70. The boiling point of water on the Kelvin scale is closest to:
    (Easy) (Domain IV Skill 1)

    A. 112 K

    B. 212 K

    C. 373 K

    D. 473 K

71. A calorie is the amount of heat energy that will:
    (Easy) (Domain IV Skill 1)

    A. Raise the temperature of one gram of water from 14.5° C to 15.5° C.

    B. Lower the temperature of one gram of water from 16.5° C to 15.5° C

    C. Raise the temperature of one gram of water from 32° F to 33° F

    D. Cause water to boil at two atmospheres of pressure.

72. A temperature change of 40 degrees Celsius is equal to a change in Fahrenheit degrees of:
    (Average Rigor) (Domain IV Skill 1)

    A. 40

    B. 20

    C. 72

    D. 112

73. Solids expand when heated because:
    (Rigorous) (Domain IV Skill 2)

    A. Molecular motion causes expansion

    B. PV = nRT

    C. Magnetic forces stretch the chemical bonds

    D. All material is effectively fluid

74. The number of calories required to raise the temperature of 40 grams of water at 30°C to steam at 100°C is:
    (Rigorous) (Domain IV Skill 4)

    A. 7500

    B. 23,000

    C. 24,400

    D. 30,500

75. Use the information on heats below to solve this problem. An ice block at 0° Celsius is dropped into 100 g of liquid water at 18° Celsius. When thermal equilibrium is achieved, only liquid water at 0° Celsius is left. What was the mass, in grams, of the original block of ice?

Given: Heat of fusion of ice = 80 cal/g
Heat of vaporization of ice = 540 cal/g
Specific Heat of ice = 0.50 cal/g°C
Specific Heat of water = 1 cal/g°C
(Rigorous) (Domain IV Skill 4)

A. 2.0

B. 5.0

C. 10.0

D. 22.5

76. Heat transfer by electromagnetic waves is termed:
(Easy) (Domain IV Skill 5)

A. Conduction

B. Convection

C. Radiation

D. Phase Change

77. A long copper bar has a temperature of 60°C at one end and 0°C at the other. The bar reaches thermal equilibrium (barring outside influences) by the process of heat:
(Average Rigor) (Domain IV Skill 5)

A. Fusion

B. Convection

C. Conduction

D. Microwaving

78. The First Law of Thermodynamics takes the form dU = dW when the conditions are:
(Rigorous) (Domain IV Skill 7)

A. Isobaric

B. Isochloremic

C. Isothermal

D. Adiabatic

79. **What is temperature?**
    (Average Rigor) (Domain IV Skill 10)

    A. Temperature is a measure of the conductivity of the atoms or molecules in a material

    B. Temperature is a measure of the kinetic energy of the atoms or molecules in a material

    C. Temperature is a measure of the relativistic mass of the atoms or molecules in a material

    D. Temperature is a measure of the angular momentum of electrons in a material

80. **Bohr's theory of the atom was the first to quantize:**
    (Average Rigor) (Domain V Skill 1)

    A. Work

    B. Angular Momentum

    C. Torque

    D. Duality

81. **Two neutral isotopes of a chemical element have the same numbers of:**
    (Easy) (Domain V Skill 2)

    A. Electrons and Neutrons

    B. Electrons and Protons

    C. Protons and Neutrons

    D. Electrons, Neutrons, and Protons

82. **Which statement best describes why population inversion is necessary for a laser to operate?**
    (Rigorous) (Domain V Skill 2)

    A. Population inversion prevents too many electrons from being excited into higher energy levels, thus preventing damage to the gain medium.

    B. Population inversion maintains a sufficient number of electrons in a higher energy state so as to allow a significant amount of stimulated emission.

    C. Population inversion prevents the laser from producing coherent light.

    D. Population inversion is not necessary for the operation of most lasers.

83. When a radioactive material emits an alpha particle only, its atomic number will:
(Average Rigor) (Domain V Skill 3)

   A. Decrease

   B. Increase

   C. Remain unchanged

   D. Change randomly

84. Ten grams of a sample of a radioactive material (half-life = 12 days) were stored for 48 days and re-weighed. The new mass of material was:
(Rigorous) (Domain V Skill 3)

   A. 1.25 g

   B. 2.5 g

   C. 0.83 g

   D. 0.625 g

85. Which of the following pairs of elements are not found to fuse in the centers of stars?
(Average Rigor) (Domain V Skill 7)

   A. Oxygen and Helium

   B. Carbon and Hydrogen

   C. Beryllium and Helium

   D. Cobalt and Hydrogen

86. The constant of proportionality between the energy and the frequency of electromagnetic radiation is known as the:
(Easy) (Domain V Skill 8)

   A. Rydberg constant

   B. Energy constant

   C. Planck constant

   D. Einstein constant

87. A crew is on-board a spaceship, traveling at 60% of the speed of light with respect to the earth. The crew measures the length of their ship to be 240 meters. When a ground-based crew measures the apparent length of the ship, it equals:
(Rigorous) (Domain V Skill 10)

   A. 400 m

   B. 300 m

   C. 240 m

   D. 192 m

88. If an object of length 5 meters (along the $\hat{y}$ axis) is traveling at 99% of the speed of light in the $\hat{x}$ direction, what is its length, as measured by an observer at the origin?
(Rigorous) (Domain V Skill 10)

   A. 0 m

   B. 0.71 m

   C. 3.4 m

   D. 5 m

89. Which statement best describes a valid approach to testing a scientific hypothesis?
(Easy) (Domain VI Skill 1)

   A. Use computer simulations to verify the hypothesis

   B. Perform a mathematical analysis of the hypothesis

   C. Design experiments to test the hypothesis

   D. All of the above

90. Which of the following is not a key purpose for the use of open communication about and peer-review of the results of scientific investigations?
    (Average Rigor) (Domain VI Skill 2)

    A. Testing, by other scientists, of the results of an investigation for the purpose of refuting any evidence contrary to an established theory

    B. Testing, by other scientists, of the results of an investigation for the purpose of finding or eliminating any errors in reasoning or measurement

    C. Maintaining an open, public process to better promote honesty and integrity in science

    D. Provide a forum to help promote progress through mutual sharing and review of the results of investigations

91. Which description best describes the role of a scientific model of a physical phenomenon?
    (Average Rigor) (Domain VI Skill 3)

    A. An explanation that provides a reasonably accurate approximation of the phenomenon

    B. A theoretical explanation that describes exactly what is taking place

    C. A purely mathematical formulation of the phenomenon

    D. A predictive tool that has no interest in what is actually occurring

92. Which of the following aspects of the use of computers for collecting experimental data is not a concern for the scientist? *(Rigorous) (Domain VI Skill 4)*

   A. The relative speeds of the processor, peripheral, memory storage unit and any other components included in data acquisition equipment

   B. The financial cost of the equipment, utilities and maintenance

   C. Numerical error due to a lack of infinite precision in digital equipment

   D. The order of complexity of data analysis algorithms

93. What best describes an appropriate view of scale in scientific investigation? *(Rigorous) (Domain VI Skill 5)*

   A. Scale is irrelevant: the same fundamental principles apply at all scale levels

   B. Scale has no bearing on experimentation, although it may have a part in some theories

   C. Scale has only a practical effect: it may make certain experiments more or less difficult due to limitations on equipment, but it has no fundamental or theoretical impact

   D. Scale is an important consideration both experimentally and theoretically: certain phenomena may have negligible effect at one scale, but may have an overwhelming effect at another scale

94. Which phenomenon was first explained using the concept of quantization of energy, thus providing one of the key foundational principles for the later development of quantum theory?
(Rigorous) (Domain VI Skill 6)

   A. The photoelectric effect

   B. Time dilation

   C. Blackbody radiation

   D. Magnetism

95. Which of the following is *not* an SI unit?
(Average Rigor) (Domain VI Skill 8)

   A. Joule

   B. Coulomb

   C. Newton

   D. Erg

96. Which statement best describes how significant figures must be treated when analyzing a collection of measured data?
(Rigorous) (Domain VI Skill 9)

   A. Report all data and any results calculated using that data with the number of significant figures that corresponds to the least accurate number (i.e., the one with the fewest significant figures)

   B. Report all data and any results calculated using that data with the number of significant figures that corresponds to the most accurate number (i.e., the one with the most significant figures)

   C. Follow appropriate rules for the propagation of error, maintaining an appropriate number of significant figures for each particular set of data or calculated result

   D. Report all data with the appropriate number of significant figures, but ignore error when calculating results

97. Which of the following best describes the relationship of precision and accuracy in scientific measurements? *(Easy) (Domain VI Skill 11)*

   A. Accuracy is how well a particular measurement agrees with the value of the actual parameter being measured; precision is how well a particular measurement agrees with the average of other measurements taken for the same value

   B. Precision is how well a particular measurement agrees with the value of the actual parameter being measured; accuracy is how well a particular measurement agrees with the average of other measurements taken for the same value

   C. Accuracy is the same as precision

   D. Accuracy is a measure of numerical error; precision is a measure of human error

98. Which of the following experiments presents the most likely cause for concern about laboratory safety? *(Average Rigor) (Domain VI Skill 13)*

   A. Computer simulation of a nuclear reactor

   B. Vibration measurement with a laser

   C. Measurement of fluorescent light intensity with a battery-powered photodiode circuit

   D. Ambient indoor ionizing radiation measurement with a Geiger counter.

99. If a particular experimental observation contradicts a theory, what is the most appropriate approach that a physicist should take? *(Average Rigor) (Domain VI Skill 22)*

   A. Immediately reject the theory and begin developing a new theory that better fits the observed results

   B. Report the experimental result in the literature without further ado

   C. Repeat the observations and check the experimental apparatus for any potential faulty components or human error, and then compare the results once more with the theory

   D. Immediately reject the observation as in error due to its conflict with theory

100. Which situation calls might best be described as involving an ethical dilemma for a scientist? *(Rigorous) (Domain VI Skill 22)*

   A. Submission to a peer-review journal of a paper that refutes an established theory

   B. Synthesis of a new radioactive isotope of an element

   C. Use of a computer for modeling a newly-constructed nuclear reactor

   D. Use of a pen-and-paper approach to a difficult problem

Answer Key

1. C	22. D	43. C	64. A	85. D
2. B	23. B	44. B	65. B	86. C
3. A	24. C	45. D	66. C	87. D
4. A	25. A	46. A	67. B	88. D
5. A	26. B	47. B	68. D	89. D
6. B	27. D	48. A	69. C	90. A
7. C	28. B	49. D	70. C	91. A
8. C	29. D	50. C	71. A	92. D
9. A	30. D	51. A	72. C	93. D
10. D	31. B	52. B	73. A	94. C
11. B	32. A	53. C	74. C	95. D
12. C	33. B	54. A	75. D	96. C
13. B	34. B	55. D	76. C	97. A
14. D	35. D	56. B	77. C	98. B
15. C	36. B	57. A	78. D	99. C
16. D	37. C	58. B	79. B	100. B
17. D	38. A	59. B	80. B	
18. C	39. A	60. A	81. B	
19. C	40. C	61. B	82. B	
20. C	41. B	62. D	83. A	
21. B	42. C	63. B	84. D	

Rigor Analysis Table

Easy	19%	3,4,8,11,17,34,35,46,48,52,55,62,70,71,76,81,86,89,97
Average Rigor	40%	1,6,7,9,13,14,15,18,19,20,25,26,29,31,36,37,39,42,43, 45,49,50,53,54,56,59,60,61,63,72,77,79,80,83,85,90, 91,95,98,99
Rigorous	41%	2,5,10,12,16,21,22,23,24,27,28,30,32,33,38,40,41,44, 47,51,57,58,64,65,66,67,68,69,73,74,75,78,82,84,87, 88,92,93,94,96,100

Rationales with Sample Questions

1. **When acceleration is plotted versus time, the area under the graph represents:**
   (Average Rigor) (Domain I Skill 3)

   A. moment in time

   B. Distance

   C. Velocity

   D. Acceleration

Answer: C

The area under a graph will have units equal to the product of the units of the two axes. (To visualize this, picture a graphed rectangle with its area equal to length times width.)
Therefore, multiply units of acceleration by units of time:
(length/time^2)(time)
This equals length/time, i.e. units of velocity.

2. **A skateboarder accelerates down a ramp, with constant acceleration of two meters per second squared, from rest. The distance in meters, covered after four seconds, is:**
   (Rigorous) (Domain I Skill 3)

   A. 10

   B. 16

   C. 23

   D. 37

Answer: B

To answer this question, recall the equation relating constant acceleration to distance and time: $x = \frac{1}{2} a t^2 + v_0 t + x_0$ where x is position; a is acceleration; t is time; v_0 and x_0 are initial velocity and position (both zero in this case). Thus, to solve for x: $x = \frac{1}{2} (2 \text{ m/s}^2)(4^2 \text{s}^2) + 0 + 0$; x = 16 m. This is consistent only with answer (B).

3. **A brick and hammer fall from a ledge at the same time. They would be expected to:**
   (Easy) (Domain I Skill 4)

   A. Reach the ground at the same time

   B. Accelerate at different rates due to difference in weight

   C. Accelerate at different rates due to difference in potential energy

   D. Accelerate at different rates due to difference in kinetic energy

Answer: A

This is a classic question about falling in a gravitational field. All objects are acted upon equally by gravity, so they should reach the ground at the same time. (In real life, air resistance can make a difference, but not at small heights for similarly shaped objects.) In any case, weight, potential energy, and kinetic energy do not affect gravitational acceleration. Thus, the only possible answer is (A).

4. **Gravitational force at the earth's surface causes:**
   (Easy) (Domain I Skill 4)

   A. All objects to fall with equal acceleration, ignoring air resistance

   B. Some objects to fall with constant velocity, ignoring air resistance

   C. A kilogram of feathers to float at a given distance above the earth

   D. Aerodynamic objects to accelerate at an increasing rate

Answer: A

Gravity acts to cause equal acceleration on all objects, though our atmosphere causes air resistance that slows some objects more than others. This is consistent only with answer (A). Answer (B) is incorrect, because ignoring air resistance leads to the result of constant acceleration, not zero acceleration. Answer (C) is incorrect because all objects (except tiny ones in which random Brownian motion is more significant than gravity) eventually fall due to gravity. Answer (D) is incorrect because it is not related to the constant acceleration due to gravity.

5. A baseball is thrown with an initial velocity of 30 m/s at an angle of 45°. Neglecting air resistance, how far away will the ball land? *(Rigorous)(Domain I Skill 4)*

      A. 92 m

      B. 78 m

      C. 65 m

      D. 46 m

Answer: A

To answer this question, recall the equations for projectile motion:
$y = \frac{1}{2} a t^2 + v_{0y} t + y_0$
$x = v_{0x} t + x_0$
where x and y are horizontal and vertical position, respectively; t is time; a is acceleration due to gravity; v_{0x} and v_{0y} are initial horizontal and vertical velocity, respectively; x_0 and y_0 are initial horizontal and vertical position, respectively.
For our case:
x_0 and y_0 can be set to zero
both v_{0x} and v_{0y} are (using trigonometry) = $(\sqrt{2}/2)$ 30 m/s
$a = -9.81$ m/s^2

We then use the vertical motion equation to find the time aloft (setting y equal to zero to find the solution for t):
$0 = \frac{1}{2} (-9.81 \text{ m/s}^2) t^2 + (\sqrt{2}/2)$ 30 m/s t
Then solving, we find:
t = 0 s (initial set-up) or t = 4.324 s (time to go up and down)

Using t = 4.324 s in the horizontal motion equation, we find:
$x = ((\sqrt{2}/2)$ 30 m/s) (4.324 s)
x = 91.71 m

This is consistent only with answer (A).

6. In order to switch between two different reference frames in special relativity, we use the _____ transformation.
   (Average Rigor) (Domain I Skill 5)

   A. Galilean

   B. Lorentz

   C. Euclidean

   D. Laplace

Answer: B

The Lorentz transformation is the set of equations to scale length and time between inertial reference frames in special relativity, when velocities are close to the speed of light.

The Galilean transformation is a parallel set of equations, used for 'classical' situations when velocities are much slower than the speed of light. Euclidean geometry is useful in physics, but not relevant here. Laplace transforms are a method of solving differential equations by using exponential functions. The correct answer is therefore (B).

7. **A mass is moving at constant speed in a circular path. Choose the true statement below:**
   (Average Rigor) (Domain I Skill 6)

   A. Two forces in equilibrium are acting on the mass.

   B. No forces are acting on the mass.

   C. One centripetal force is acting on the mass.

   D. One force tangent to the circle is acting on the mass.

Answer: C

To answer this question, recall that by Newton's 2nd Law, $F = ma$. In other words, force is mass times acceleration. Furthermore, acceleration is any change in the velocity vector—whether in size or direction. In circular motion, the direction of velocity is constantly changing. Therefore, there must be an unbalanced force on the mass to cause that acceleration. This eliminates answers (A) and (B) as possibilities. Recall then that the mass would ordinarily continue traveling tangent to the circle (by Newton's 1st Law). Therefore, the force must be to cause the turn, i.e. a centripetal force. Thus, the answer can only be (C).

8. **The magnitude of a force is:**
   (Easy) (Domain I Skill 6)

   A. Directly proportional to mass and inversely to acceleration

   B. Inversely proportional to mass and directly to acceleration

   C. Directly proportional to both mass and acceleration

   D. Inversely proportional to both mass and acceleration

Answer: C

To solve this problem, recall Newton's 2nd Law, i.e. net force is equal to mass times acceleration. Therefore, the only possible answer is (C).

9. If a force of magnitude *F* gives a mass *M* an acceleration *A*, then a force 3*F* would give a mass 3*M* an acceleration:
(Average Rigor) (Domain I Skill 6)

    A. *A*

    B. 12*A*

    C. *A*/2

    D. 6*A*

Answer: A

To solve this problem, apply Newton's Second Law, which is also implied by the first part of the problem:
Force = (Mass)(Acceleration)
F = MA
Then apply the same law to the second case, and isolate the unknown:
3*F* = 3*M* x
x = (3*F*)/(3*M*)
x = *F/M*
x = *A* (by substituting from our first equation)
Only answer (A) matches these calculations.

10. An inclined plane is tilted by gradually increasing the angle of elevation θ, until the block will slide down at a constant velocity. The coefficient of friction, μ_k, is given by:
   (Rigorous) (Domain I Skill 7)

   A. cos θ

   B. sin θ

   C. cosecant θ

   D. tangent θ

Answer: D

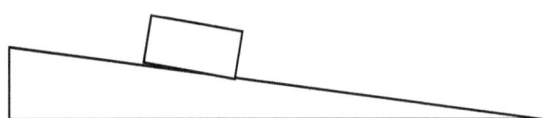

When the block moves, its force upstream (due to friction) must equal its force downstream (due to gravity).

The friction force is given by
$F_f = \mu_k N$
where μ_k is the friction coefficient and N is the normal force.

Using similar triangles, the gravity force is given by
$F_g = mg \sin \theta$
and the normal force is given by
$N = mg \cos \theta$
When the block moves at constant velocity, it must have zero net force, so set equal the force of gravity and the force due to friction:
$F_f = F_g$
$\mu_k mg \cos \theta = mg \sin \theta$
$\mu_k = \tan \theta$

Answer (D) is the only appropriate choice in this case.

11. A classroom demonstration shows a needle floating in a tray of water. This demonstrates the property of:
 (Easy) (Domain I Skill 8)

    A. Specific Heat

    B. Surface Tension

    C. Oil-Water Interference

    D. Archimedes' Principle

Answer: B

To answer this question, note that the only information given is that the needle (a small object) floats on the water. This occurs because although the needle is denser than the water, the surface tension of the water causes sufficient resistance to support the small needle. Thus the answer can only be (B). Answer (A) is unrelated to objects floating, and while answers (C) and (D) could be related to water experiments, they are not correct in this case. There is no oil in the experiment, and Archimedes' Principle allows the equivalence of displaced volumes, which is not relevant here.

12. A uniform pole weighing 100 grams, that is one meter in length, is supported by a pivot at 40 centimeters from the left end. In order to maintain static position, a 200 gram mass must be placed _____ centimeters from the left end.
   (Rigorous) (Domain I Skill 9)

   A. 10

   B. 45

   C. 35

   D. 50

Answer: C

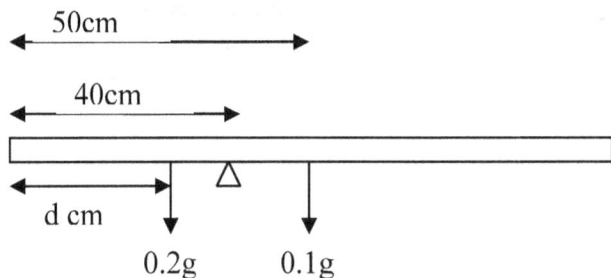

Since the pole is uniform, we can assume that its weight 0.1g acts at the center, i.e. 50 cm from the left end. In order to keep the pole balanced on the pivot, the 200 gram mass must be placed such that the torque on the pole due to the mass is equal and opposite to the torque due to the pole's weight. Thus, if the 200 gram mass is placed d cm from the left end of the pole,

$(40 - d) \times 0.2g = 10 \times 0.1g$; $40 - d = 5$; $d = 35$ cm

13. A satellite is in a circular orbit above the earth. Which statement is false?
    (Average Rigor) (Domain I Skill 10)

    A. An external force causes the satellite to maintain orbit.

    B. The satellite's inertia causes it to maintain orbit.

    C. The satellite is accelerating toward the earth.

    D. The satellite's velocity and acceleration are not in the same direction.

Answer: B

To answer this question, recall that in circular motion, an object's inertia tends to keep it moving straight (tangent to the orbit), so a centripetal force (leading to centripetal acceleration) must be applied. In this case, the centripetal force is gravity due to the earth, which keeps the object in motion. Thus, (A), (C), and (D) are true, and (B) is the only false statement.

14. A 100 g mass revolving around a fixed point, on the end of a 0.5 meter string, circles once every 0.25 seconds. What is the magnitude of the centripetal acceleration?
    (Average Rigor) (Domain I Skill 10)

    A. 1.23 m/s^2

    B. 31.6 m/s^2

    C. 100 m/s^2

    D. 316 m/s^2

Answer: D

The centripetal acceleration is equal to the product of the radius and the square of the angular frequency ω. In this case, ω is equal to 25.1 Hz. Squaring this value and multiplying by 0.5 m yields the result in answer D.

TEACHER CERTIFICATION STUDY GUIDE

15. The kinetic energy of an object is _____ proportional to its _____.
    (Average Rigor) (Domain I Skill 11)

    A. Inversely…inertia

    B. Inversely…velocity

    C. Directly…mass

    D. Directly…time

Answer: C

To answer this question, recall that kinetic energy is equal to one-half of the product of an object's mass and the square of its velocity:
$KE = \frac{1}{2} m v^2$

Therefore, kinetic energy is directly proportional to mass, and the answer is (C). Note that although kinetic energy is associated with both velocity and momentum (a measure of inertia), it is not *inversely* proportional to either one.

16. A force is given by the vector 5 N x + 3 N y (where x and y are the unit vectors for the x- and y- axes, respectively). This force is applied to move a 10 kg object 5 m, in the x direction. How much work was done?
    (Rigorous) (Domain I Skill 11)

    E. 250 J

    F. 400 J

    G. 40 J

    H. 25 J

Answer: D

To find out how much work was done, note that work counts only the force in the direction of motion. Therefore, the only part of the vector that we use is the 5 N in the x-direction. Note, too, that the mass of the object is not relevant in this problem. We use the work equation:
Work = (Force in direction of motion) (Distance moved)
Work = (5 N) (5 m) = 25 J.
This is consistent only with answer (D).

TEACHER CERTIFICATION STUDY GUIDE

17. **An object traveling through air loses part of its energy of motion due to friction. Which statement best describes what has happened to this energy?**
    (Easy) (Domain I Skill 12)

    A. The energy is destroyed

    B. The energy is converted to static charge

    C. The energy is radiated as electromagnetic waves

    D. The energy is lost to heating of the air

Answer: D

Since energy must be conserved, the energy of motion of the object is converted, in part, to energy of motion of the molecules in the air (and, to some extent, in the object). This additional motion is equivalent to an increase in heat. Thus, friction is a loss of energy of motion through heating.

18. **If the internal energy of a system remains constant, how much work is done by the system if 1 kJ of heat energy is added?**
    (Average Rigor) (Domain I Skill 12)

    A. 0 kJ

    B. -1 kJ

    C. 1 kJ

    D. 3.14 kJ

Answer: C

According to the first law of thermodynamics, if the internal energy of a system remains constant, then any heat energy added to the system must be balanced by the system performing work on its surroundings. In the case of an ideal gas, the gas would necessarily expand when heated, assuming a constant internal energy was somehow maintained. Applying conservation of energy, answer C is found to be correct.

19. **A mass of 2 kg connected to a spring undergoes simple harmonic motion at a frequency of 3 Hz. What is the spring constant?**
(Average Rigor) (Domain I Skill 13)

    A. 6 kg/s^2

    B. 18 kg/s^2

    C. 710 kg/s^2

    D. 1000 kg/s^2

Answer: C

The spring constant, k, is equal to $m\omega^2$. In this case, ω is equal to 2π times the frequency of 3 Hz. The spring constant may be derived quickly by recognizing that the position of the mass varies sinusoidally with time at an angular frequency ω. Noting that the acceleration is the second derivative of the position with respect to time, the expression for k in Hooke's law (F = -kx) can be easily derived.

20. **Which statement best describes the relationship of simple harmonic motion to a simple pendulum of length L, mass m and displacement of arc length s?**
    (Average Rigor) (Domain I Skill 13)

    A. A simple pendulum cannot be modeled using simple harmonic motion

    B. A simple pendulum may be modeled using the same expression as Hooke's law for displacement s, but with a spring constant equal to the tension on the string

    C. A simple pendulum may be modeled using the same expression as Hooke's law for displacement s, but with a spring constant equal to m g/L

    D. A simple pendulum typically does not undergo simple harmonic motion

Answer: C

The force on a simple pendulum may be expressed approximately (when displacement s is small) according to the following equation:

$$F \approx -\frac{mg}{L}s$$

This expression has the same form as Hooke's law (F = -kx). Thus, answer C is the most correct response. Another approach to the question is to eliminate answers A and D as obviously incorrect, and then to eliminate answer B as not having appropriate units for the spring constant.

TEACHER CERTIFICATION STUDY GUIDE

21. What is the maximum displacement from equilibrium of a 1 kg mass that is attached to a spring with constant k = 100 kg/s^2 if the mass has a velocity of 3 m/s at the equilibrium point?
    (Rigorous) (Domain I Skill 13)

    A. 0.1 m

    B. 0.3 m

    C. 3 m

    D. 10 m

Answer: B

This problem may be solved using simple differential equation analysis. Alternatively, it may be noted that since harmonic motion involves sinusoidal functions, one may simply assume that the mass has a displacement x(t) = A sin ωt. Thus, at time t = 0, the mass is at the equilibrium point (x = 0). The spring constant may be used with the known mass to determine the frequency of oscillation according to the equation k = mω^2. The result is ω = 10 radians/sec. Differentiating x(t) with respect to time and setting the result equal to 3 m/s at time t = 0 allows for the determination of A, which is 0.3 m. This is the maximum displacement of the mass.

Alternatively, recalling the expression for the potential energy of a spring, we can use conservation of energy and set the total energy at the equilibrium point (entirely kinetic) equal to the total energy at the point of maximum displacement (entirely potential): 1/2mv^2=1/2kA^2.

22. Which of the following is not an assumption upon which the kinetic-molecular theory of gases is based?
    (Rigorous) (Domain I Skill 14)

    A. Quantum mechanical effects may be neglected

    B. The particles of a gas may be treated statistically

    C. The particles of the gas are treated as very small masses

    D. Collisions between gas particles and container walls are inelastic

Answer: D

Since the kinetic-molecular theory is classical in nature, quantum mechanical effects are indeed ignored, and answer A is incorrect. The theory also treats gases as a statistical collection of point-like particles with finite masses. As a result, answers B and C may also be eliminated. Thus, answer D is correct: collisions between gas particles and container walls are treated as elastic in the kinetic-molecular theory.

23. A projectile with a mass of 1.0 kg has a muzzle velocity of 1500.0 m/s when it is fired from a cannon with a mass of 500.0 kg. If the cannon slides on a frictionless track, it will recoil with a velocity of ____ m/s.
(Rigorous) (Domain I Skill 14)

    A. 2.4

    B. 3.0

    C. 3.5

    D. 1500

Answer: B

To solve this problem, apply Conservation of Momentum to the cannon-projectile system. The system is initially at rest, with total momentum of 0 kg m/s. Since the cannon slides on a frictionless track, we can assume that the net momentum stays the same for the system. Therefore, the momentum forward (of the projectile) must equal the momentum backward (of the cannon). Thus:

$p_{projectile} = p_{cannon}$
$m_{projectile} V_{projectile} = m_{cannon} V_{cannon}$
$(1.0 \text{ kg})(1500.0 \text{ m/s}) = (500.0 \text{ kg})(x)$
$x = 3.0 \text{ m/s}$
Only answer (B) matches these calculations.

24. A car (mass m_1) is driving at velocity v, when it smashes into an unmoving car (mass m_2), locking bumpers. Both cars move together at the same velocity. The common velocity will be given by: *(Rigorous) (Domain I Skill 14)*

   A. m_1v/m_2

   B. m_2v/m_1

   C. $m_1v/(m_1 + m_2)$

   D. $(m_1 + m_2)v/m_1$

Answer: C

In this problem, there is an inelastic collision, so the best method is to assume that momentum is conserved. (Recall that momentum is equal to the product of mass and velocity.)
Therefore, apply Conservation of Momentum to the two-car system:
Momentum at Start = Momentum at End
(Mom. of Car 1) + (Mom. of Car 2) = (Mom. of 2 Cars Coupled)
$m_1v + 0 = (m_1 + m_2)x$
$x = m_1v/(m_1 + m_2)$
Only answer (C) matches these calculations.

Watch out for the other answers, because errors in algebra could lead to a match with incorrect answer (D), and assumption of an elastic collision could lead to a match with incorrect answer (A).

25. Which of the following units is not used to measure torque?
    (Average Rigor) (Domain I Skill 15)

    A. slug ft

    B. lb ft

    C. N m

    D. dyne cm

Answer: A

To answer this question, recall that torque is always calculated by multiplying units of force by units of distance. Therefore, answer (A), which is the product of units of mass and units of distance, must be the choice of incorrect units. Indeed, the other three answers all could measure torque, since they are of the correct form. It is a good idea to review "English Units" before the teacher test, because they are occasionally used in problems.

26. Which statement best describes an approach for calculating the kinetic energy of a rotating object?
    (Average Rigor) (Domain I Skill 15)

    A. Rotating objects have no kinetic energy; only objects that undergo linear motion have kinetic energy

    B. Treat the object as a collection of small unit volumes (or particles) and sum the kinetic energies of all the constituent parts

    C. Calculate the rotational inertia, which is equal to the kinetic energy

    D. The kinetic energy of a rotating object cannot be calculated

Answer: B

An object undergoing rotation has a kinetic energy that can indeed be calculated. Thus, answers A and D may be discarded. Since rotational inertia is not equal to an energy value, answer B may be accepted by elimination of the alternatives. The kinetic energy of a rotating object is the sum of the kinetic energies of the set of constituent parts of the object, given the index i. The equation below is an expression of this approach.

$$KE = \sum_i \frac{1}{2} m_i v_i^2$$

27. Given the following values for the masses of a proton, a neutron and an alpha particle, what is the nuclear binding energy of an alpha particle?
(Rigorous) (Domain I Skill 16)

Proton mass = 1.6726×10^{-27} kg
Neutron mass = 1.6749×10^{-27} kg
Alpha particle mass = 6.6465×10^{-27} kg

    A. 0 J

    B. 7.3417×10^{-27} J

    C. 4 J

    D. 4.3589×10^{-12} J

Answer: D

The nuclear binding energy is the amount of energy that is required to break the nucleus into its component nucleons. In this case, the binding energy of an alpha particle, which is composed of two protons and two neutrons, is calculated by first finding the difference between the sum of the masses of all the nucleons and the mass of the alpha particle. Using the equation $E = mc^2$ to find the energy in terms of the mass difference of 4.85×10^{-29} kg, and using the speed of light of about 2.9979×10^8 m/s, the result is the value given in answer D.

28. The weight of an object on the earth's surface is designated x. When it is two earth's radii from the surface of the earth, its weight will be: (Rigorous) (Domain I Skill 17)

    A. $x/4$

    B. $x/9$

    C. $4x$

    D. $16x$

Answer: B

To solve this problem, apply the universal Law of Gravitation to the object and Earth:
$F_{gravity} = (GM_1M_2)/R^2$
Because the force of gravity varies with the square of the radius between the objects, the force (or weight) on the object will be decreased by the square of the multiplication factor on the radius. Note that the object on Earth's surface is *already* at one radius from Earth's center. Thus, when it is two radii from Earth's surface, it is three radii from Earth's center. R^2 is then nine, so the weight is $x/9$. Only answer (B) matches these calculations.

29. Given a vase full of water, with holes punched at various heights. The water squirts out of the holes, achieving different distances before hitting the ground. Which of the following accurately describes the situation?
(Average Rigor) (Domain I Skill 20)

    A. Water from higher holes goes farther, due to Pascal's Principle.

    B. Water from higher holes goes farther, due to Bernoulli's Principle.

    C. Water from lower holes goes farther, due to Pascal's Principle.

    D. Water from lower holes goes farther, due to Bernoulli's Principle.

Answer: D

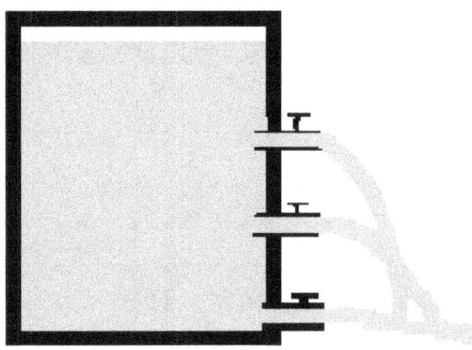

To answer this question, consider the pressure on the water in the vase. The deeper the water, the higher the pressure. Thus, when a hole is punched, the water stream will achieve higher velocity as it equalizes to atmospheric pressure. The lower streams will therefore travel farther before hitting the ground. This eliminates answers (A) and (B). Then recall that Pascal's Principle provides for immediate pressure changes throughout a fluid, while Bernoulli's Principle translates pressure, velocity, and height energy into each other. In this case, the pressure energy is being transformed into velocity energy, and Bernoulli's Principle applies. Therefore, the only appropriate answer is (D).

30. The electric force in Newtons, on two small objects (each charged to –10 microCoulombs and separated by 2 meters) is:
(Rigorous) (Domain II Skill 1)

    A. 1.0

    B. 9.81

    C. 31.0

    D. 0.225

Answer: D

To answer this question, use Coulomb's Law, which gives the electric force between two charged particles:
$F = k Q_1 Q_2 / r^2$
Then our unknown is F, and our knowns are:
$k = 9.0 \times 10^9$ Nm^2/C^2
$Q_1 = Q_2 = -10 \times 10^{-6}$ C
$r = 2$ m

Therefore
$F = (9.0 \times 10^9)(-10 \times 10^{-6})(-10 \times 10^{-6})/(2^2)$ N
$F = 0.225$ N

This is compatible only with answer (D).

31. Which of the following is a legitimate explanation for lightning? *(Average Rigor) (Domain II Skill 2)*

    A. Lightning is the result of varying magnetic fields in clouds

    B. Lightning is the result of an electric potential difference greater than the breakdown voltage of air

    C. Lightning is the result of a lens effect due to air masses with different temperatures

    D. Lightning is the result of global warming

Answer: B

When an electric charge differential is formed in the atmosphere, or between some portion of the atmosphere and the ground, an electric potential is formed. When this electric potential exceeds the breakdown voltage of air, the molecules that make up the air are ionized, thus forming a conducting path for an electrical discharge. This discharge is seen as lightning.

32. A hollow conducting sphere of radius R is charged with a total charge Q. What is the magnitude of the electric field at a distance r (given r<R) from the center of the sphere? (k is the electrostatic constant)
   (Rigorous) (Domain II Skill 2)

   A. 0

   B. $k Q/R^2$

   C. $k Q/(R^2 - r^2)$

   D. $k Q/(R - r)^2$

Answer: A

You may be tempted to use the equation for the electric field:
$E = F/Q$ (E = electric field; F = electric force; Q = charge)

and the Coulomb's Law expression for electric force:
$F = k Q_1 Q_2 / r^2$ (k = constant; Q_1 and Q_2 = charges; r = distance apart),

which usually would give
$E = k Q/(R-r)^2$ in a similar context.

However, this question addresses a special case, i.e. a hollow conductor. Inside a hollow conductor, no electric field exists, because if there were an electric field inside, the conductor's free electrons would be forced to move (by the electric force) until the electric field became zero. Therefore, the only correct answer is (A).

33. What is the electric flux density (or electric displacement) through each face of a cube, of side length 2 meters, that contains a central point charge of 2 Coulombs?
    (Rigorous) (Domain II Skill 2)

    A. 0.50 Coulomb/m^2

    B. 0.33 Coulomb/m^2

    C. 0.13 Coulomb/m^2

    D. 0.33 Tesla

Answer: B

Since the point charge is located at the center of the cube, a symmetric situation exists. Furthermore, since electric flux density has the same value regardless of the shape or size of the surface that encloses the charge, the size of the cube is irrelevant. The amount of flux, therefore, is the same through each face of the cube. Simply divide the charge by the number of cube faces (six) to obtain the result in answer B.

34. Static electricity generation occurs by:
    (Easy) (Domain II Skill 3)

    A. Telepathy

    B. Friction

    C. Removal of heat

    D. Evaporation

Answer: B

Static electricity occurs because of friction and electric charge build-up. There is no such thing as telepathy, and neither removal of heat nor evaporation are causes of static electricity. Therefore, the only possible answer is (B).

35. A semi-conductor allows current to flow:
(Easy) (Domain II Skill 3)

    A. Never

    B. Always

    C. As long as it stays below a maximum temperature

    D. When a minimum voltage is applied

Answer: D

To answer this question, recall that semiconductors do not conduct as well as conductors (eliminating answer (B)), but they conduct better than insulators (eliminating answer (A)). Semiconductors can conduct better when the temperature is higher (eliminating answer (C)), and their electrons move most readily under a potential difference. Thus the answer can only be (D).

36. What should be the behavior of an electroscope, which has been grounded in the presence of a positively charged object (1), after the ground connection is removed and then the charged object is removed from the vicinity (2)?
(Average Rigor) (Domain II Skill 3)

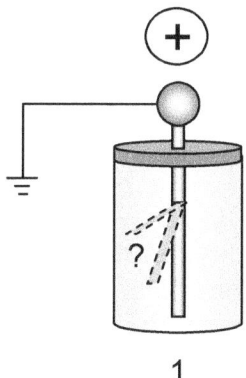
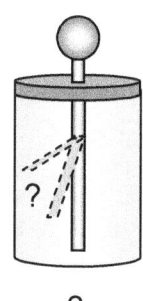

1   2

A. The metal leaf will start deflected (1) and then relax to an undeflected position (2)

B. The metal leaf will start in an undeflected position (1) and then be deflected (2)

C. The metal leaf will remain undeflected in both cases

D. The metal leaf will be deflected in both cases

Answer: B

When grounded, the electroscope will show no deflection. Nevertheless, if the ground is then removed and the charged object taken from the vicinity (in that order), the excess charge that existed near the sphere of the electroscope will distribute itself throughout the instrument, resulting in an overall net excess charge that will deflect the metal leaf.

37. All of the following use semi-conductor technology, except a(n):
    (Average Rigor) (Domain II Skill 5)

    A. Transistor

    B. Diode

    C. Capacitor

    D. Operational Amplifier

Answer: C

Semi-conductor technology is used in transistors and operational amplifiers, and diodes are the basic unit of semi-conductors. Therefore the only possible answer is (C), and indeed a capacitor does not require semi-conductor technology.

38. A 10 ohm resistor and a 50 ohm resistor are connected in parallel. If the current in the 10 ohm resistor is 5 amperes, the current (in amperes) running through the 50 ohm resistor is:
    (Rigorous) (Domain II Skill 7)

    A. 1

    B. 50

    C. 25

    D. 60

Answer: A

To answer this question, use Ohm's Law, which relates voltage to current and resistance:
$V = IR$
where V is voltage; I is current; R is resistance.

We also use the fact that in a parallel circuit, the voltage is the same across the branches.

Because we are given that in one branch, the current is 5 amperes and the resistance is 10 ohms, we deduce that the voltage in this circuit is their product, 50 volts (from $V = IR$).

We then use $V = IR$ again, this time to find I in the second branch. Because V is 50 volts, and R is 50 ohm, we calculate that I has to be 1 ampere.

This is consistent only with answer (A).

39. **When the current flowing through a fixed resistance is doubled, the amount of heat generated is:**
(Average Rigor) (Domain II Skill 9)

    A. Quadrupled

    B. Doubled

    C. Multiplied by pi

    D. Halved

Answer: A

To answer this question, recall that heat generated will occur because of the power of the circuit (power is energy per time). For a circuit with a fixed resistance:
$P = IV$
where P is power; I is current; V is voltage. Then use Ohm's Law:
$V = IR$
where V is voltage; I is current; R is resistance, and substitute:
$P = I^2 R$
and so the doubling of the current I will lead to a quadrupling of the power, and therefore the a quadrupling of the heat.

This is consistent only with answer (A). If you weren't sure of the equations, you could still deduce that with more current, there would be more heat generated, and therefore eliminate answer choice (D) in any case.

40. The greatest number of 100 watt lamps that can be connected in parallel with a 120 volt system without blowing a 5 amp fuse is: *(Rigorous) (Domain II Skill 9)*

    A. 24

    B. 12

    C. 6

    D. 1

Answer: C

To solve fuse problems, you must add together all the drawn current in the parallel branches, and make sure that it is less than the fuse's amp measure. Because we know that electrical power is equal to the product of current and voltage, we can deduce that:
$I = P/V$ (I = current (amperes); P = power (watts); V = voltage (volts))

Therefore, for each lamp, the current is 100/120 amperes, or 5/6 ampere. The highest possible number of lamps is thus six, because six lamps at 5/6 ampere each adds to 5 amperes; more will blow the fuse.

This is consistent only with answer (C).

41. The potential difference across a five Ohm resistor is five Volts. The power used by the resistor, in Watts, is:
(Rigorous) (Domain II Skill 9)

    A. 1

    B. 5

    C. 10

    D. 20

Answer: B

To answer this question, recall the two relevant equations for potential difference and electric power:
$V = IR$ (where V is voltage; I is current; R is resistance)
$P = IV = I^2R$ (where P is power; I is current; R is resistance)

Thus, first calculate the current from the first equation:
$I = V/R = 1$ Ampere

And then use the second equation:
$P = I^2R = 5$ Watts

This is consistent only with answer (B).

42. What is the resonant angular frequency (ω) of the following circuit? *(Average Rigor) (Domain II Skill 10)*

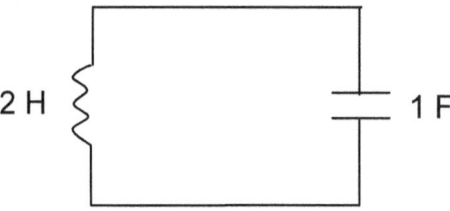

A. 1 Hz

B. 0.5 Hz

C. 0.71 radians/sec

D. This is not a resonant circuit

Answer: C

The resonant frequency of an LC circuit is found according to the following formula:

$$\omega = \sqrt{\frac{1}{LC}}$$

Here, L is in Henrys (H) and C is in Farads (F). In this case, the result of this calculation is answer C, 0.71 radians/sec. It must be noted that the angular frequency is ω (given in radians/sec), but the frequency f is ω/(2π) (given in Hz).

43. How much power is dissipated through the following resistive circuit?
    (Average Rigor) (Domain II Skill 10)

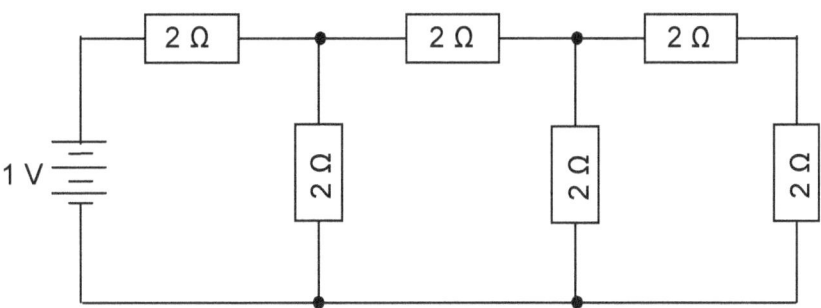

   A. 0 W

   B. 0.22 W

   C. 0.31 W

   D. 0.49 W

Answer: C

Use the rules of series and parallel resistors to quickly form an equivalent circuit with a single voltage source and a single resistor. In this case, the equivalent resistance is 3.25 Ω. The power dissipated by the circuit is the square of the voltage divided by the resistance. The final answer is C.

44. **What effect might an applied external magnetic field have on the magnetic domains of a ferromagnetic material?**
    (Rigorous) (Domain II Skill 13)

    A. The domains that are not aligned with the external field increase in size, but those that are aligned decrease in size

    B. The domains that are not aligned with the external field decrease in size, but those that are aligned increase in size

    C. The domains align perpendicular to the external field

    D. There is no effect on the magnetic domains

Answer: B

Recall that ferromagnetic domains are portions of a magnetic material that have a local magnetic moment. The material may have an overall lack of a magnetic moment due to random alignment of its domains. In the presence of an applied field, the domains may align with the field to some extent, or the boundaries of the domains may shift to give greater weight to those domains that are aligned with the field, at the expense of those domains that are not aligned with the field. As a result, of the possibilities above, B is the best answer.

45. **Which of the following statements may be taken as a legitimate inference based upon the Maxwell equation that states $\nabla \cdot \mathbf{B} = 0$?**
    (Average Rigor) (Domain II Skill 14)

    A. The electric and magnetic fields are decoupled

    B. The electric and magnetic fields are mediated by the W boson

    C. There are no photons

    D. There are no magnetic monopoles

Answer: D

Since the divergence of the magnetic flux density is always zero, there cannot be any magnetic monopoles (charges), given this Maxwell equation. If Gauss's law is applied to magnetic flux in the same manner as it is to electric flux, then the total magnetic "charge" contained within any closed surface must always be zero. This is another way of viewing the problem. Thus, answer D is correct. This answer may also be chosen by elimination of the other statements, which are untenable.

46. What is the direction of the magnetic field at the center of the loop of current (I) shown below (i.e., at point A)?
    (Easy) (Domain II Skill 15)

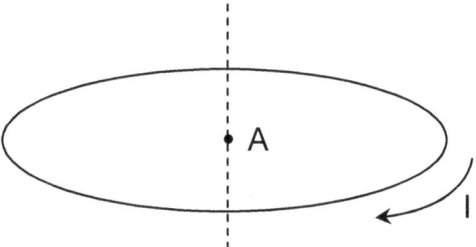

   A. Down, along the axis (dotted line)

   B. Up, along the axis (dotted line)

   C. The magnetic field is oriented in a radial direction

   D. There is no magnetic field at point A

Answer: A

The magnetic field may be found by applying the right-hand rule. The magnetic field curls around the wire in the direction of the curled fingers when the thumb is pointed in the direction of the current. Since there is a degree of symmetry, with point A lying in the center of the loop, the contributions of all the current elements on the loop must yield a field that is either directed up or down at the axis. Use of the right-hand rule indicates that the field is directed down. Thus, answer A is correct.

47. **What is the effect of running current in the same direction along two parallel wires, as shown below?**
(Rigorous) (Domain II Skill 15)

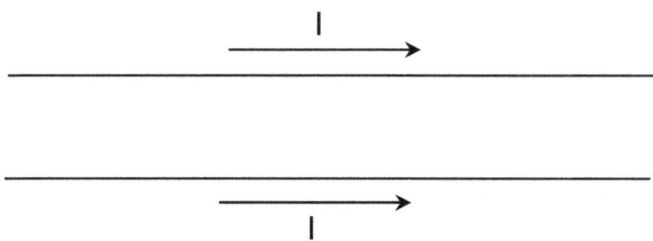

- A. There is no effect
- B. The wires attract one another
- C. The wires repel one another
- D. A torque is applied to both wires

Answer: B

Since the direction of the force on a current element is proportional to the cross product of the direction of the current element and the magnetic field, there is either an attractive or repulsive force between the two wires shown above. Using the right hand rule, it can be found that the magnetic field on the top wire due to the bottom wire is directed out of the plane of the page. Performing the cross product shows that the force on the upper wire is directed toward the lower wire. A similar argument can be used for the lower wire. Thus, the correct answer is B: an attractive force is exerted on the wires.

48. The current induced in a coil is defined by which of the following laws?
   (Easy) (Domain II Skill 16)

   E. Lenz's Law

   F. Burke's Law

   G. The Law of Spontaneous Combustion

   H. Snell's Law

Answer: A

Lenz's Law states that an induced electromagnetic force always gives rise to a current whose magnetic field opposes the original flux change. There is no relevant "Snell's Law," "Burke's Law," or "Law of Spontaneous Combustion" in electromagnetism. (In fact, only Snell's Law is a real law of these three, and it refers to refracted light.) Therefore, the only appropriate answer is (A).

49. A static magnetic flux density of 1 Tesla is linked by a wire loop, as shown below. What is the electromotive force (EMF) around the loop?
(Average Rigor) (Domain II Skill 16)

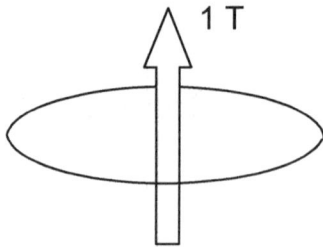

A. 1 Volt

B. 2 Volts

C. 5 Volts

D. 0 Volts

Answer: D

According to Maxwell's equations, the curl of the electric field is proportional to the time rate of change of the magnetic field.

$$\nabla \times \mathbf{E} = -\frac{\partial \mathbf{B}}{\partial t}$$

Since the situation above is source-free, the divergence of the electric field is zero. Since, as noted in the question, the magnetic flux is static, the variation of the magnetic field with time must also be zero, as must the curl of the electric field. Since the curl and divergence of the electric field are both zero, the field itself must be zero, and there can be no potential difference around the wire. Put simply, only when there is a change in the magnetic flux through the wire loop is there a corresponding EMF.

TEACHER CERTIFICATION STUDY GUIDE

50. A light bulb is connected in series with a rotating coil within a magnetic field. The brightness of the light may be increased by any of the following except:
(Average Rigor) (Domain II Skill 16)

    A. Rotating the coil more rapidly.

    B. Using more loops in the coil.

    C. Using a different color wire for the coil.

    D. Using a stronger magnetic field.

Answer: C

To answer this question, recall that the rotating coil in a magnetic field generates electric current, by Faraday's Law. Faraday's Law states that the amount of emf generated is proportional to the rate of change of magnetic flux through the loop. This increases if the coil is rotated more rapidly (A), if there are more loops (B), or if the magnetic field is stronger (D). Thus, the only answer to this question is (C).

51. The use of two circuits next to each other, with a change in current in the primary circuit, demonstrates:
(Rigorous) (Domain II Skill 16)

    A. Mutual current induction

    B. Dielectric constancy

    C. Harmonic resonance

    D. Resistance variation

Answer: A

To answer this question, recall that changing current induces a change in magnetic flux, which in turn causes a change in current to oppose that change (Lenz's and Faraday's Laws). Thus, (A) is correct. If you did not remember that, note that harmonic resonance is irrelevant here (eliminating (C)), and there is no change in resistance in the circuits (eliminating (D)).

TEACHER CERTIFICATION STUDY GUIDE

52. An electromagnetic wave propagates through a vacuum. Independent of its wavelength, it will move with constant:
    (Easy) (Domain III Skill 1)

    A. Acceleration

    B. Velocity

    C. Induction

    D. Sound

Answer: B

Electromagnetic waves are considered always to travel at the speed of light, so answer (B) is correct. Answers (C) and (D) can be eliminated in any case, because induction is not relevant here, and sound does not travel in a vacuum.

53. A wave generator is used to create a succession of waves. The rate of wave generation is one every 0.33 seconds. The period of these waves is:
    (Average Rigor) (Domain III Skill 1)

    A. 2.0 seconds

    B. 1.0 seconds

    C. 0.33 seconds

    D. 3.0 seconds

Answer: C

The definition of a period is the length of time between wave crests. Therefore, when waves are generated one per 0.33 seconds, that same time (0.33 seconds) is the period. This is consistent only with answer (C). Do not be trapped into calculating the number of waves per second, which might lead you to choose answer (D).

TEACHER CERTIFICATION STUDY GUIDE

54. **A wave has speed 60 m/s and wavelength 30,000 m. What is the frequency of the wave?**
    (Average Rigor) (Domain III Skill 1)

    A. 2.0×10^{-3} Hz

    B. 60 Hz

    C. 5.0×10^{2} Hz

    D. 1.8×10^{6} Hz

Answer: A

To answer this question, recall that wave speed is equal to the product of wavelength and frequency. Thus:
60 m/s = (30,000 m) (frequency)
frequency = 2.0×10^{-3} Hz

This is consistent only with answer (A).

55. **Rainbows are created by:**
    (Easy) (Domain III Skill 3)

    A. Reflection, dispersion, and recombination

    B. Reflection, resistance, and expansion

    C. Reflection, compression, and specific heat

    D. Reflection, refraction, and dispersion

Answer: D

To answer this question, recall that rainbows are formed by light that goes through water droplets and is dispersed into its colors. This is consistent with both answers (A) and (D). Then note that refraction is important in bending the differently colored light waves, while recombination is not a relevant concept here. Therefore, the answer is (D).

PHYSICS

56. The wave phenomenon of polarization applies only to: *(Average Rigor) (Domain III Skill 4)*

   A. Longitudinal waves

   B. Transverse waves

   C. Sound

   D. Light

Answer: B

To answer this question, recall that polarization is when waves are screened so that they come out aligned in a certain direction. (To illustrate this, take two pairs of polarizing sunglasses, and note the light differences when rotating one lens over another. When the lenses are polarizing perpendicularly, no light gets through.) This applies only to transverse waves, which have wave parts to align. Light can be polarized, but it is not the only wave that can be. Thus, the correct answer is (B).

57. A vibrating string's frequency is _____ proportional to the _____.

(Rigorous) (Domain III Skill 4)

    A. Directly; Square root of the tension

    B. Inversely; Length of the string

    C. Inversely; Squared length of the string

    D. Inversely; Force of the plectrum

Answer: A

To answer this question, recall that
$f = (n v) / (2 L)$ where f is frequency; v is velocity; L is length

and

$v = (F_{tension} / (m / L))^{1/2}$ where $F_{tension}$ is tension; m is mass; others as above

so

$f = (n / 2 L) ((F_{tension} / (m / L))^{1/2})$

indicating that frequency is directly proportional to the square root of the tension force. This is consistent only with answer (A). Note that in the final frequency equation, there is an inverse relationship with the square root of the length (after canceling like terms). This is not one of the options, however.

58. A stationary sound source produces a wave of frequency F. An observer at position A is moving toward the horn, while an observer at position B is moving away from the horn. Which of the following is true?
(Rigorous) (Domain III Skill 4)

    A. $F_A < F < F_B$

    B. $F_B < F < F_A$

    C. $F < F_A < F_B$

    D. $F_B < F_A < F$

Answer: B

To answer this question, recall the Doppler Effect. As a moving observer approaches a sound source, s/he intercepts wave fronts sooner than if s/he were standing still. Therefore, the wave fronts seem to be coming more frequently. Similarly, as an observer moves away from a sound source, the wave fronts take longer to reach him/her. Therefore, the wave fronts seem to be coming less frequently. Because of this effect, the frequency at B will seem lower than the original frequency, and the frequency at A will seem higher than the original frequency. The only answer consistent with this is (B). Note also, that even if you weren't sure of which frequency should be greater/smaller, you could still reason that A and B should have opposite effects, and be able to eliminate answer choices (C) and (D).

59. **The velocity of sound is greatest in:**
(Average Rigor) (Domain III Skill 5)

    A. Water

    B. Steel

    C. Alcohol

    D. Air

Answer: B

Sound is a longitudinal wave, which means that it shakes its medium in a way that propagates as sound traveling. The speed of sound depends on both elastic modulus and density, but for a comparison of the above choices, the answer is always that sound travels faster through a solid like steel, than through liquids or gases. Thus, the answer is (B).

60. **The combination of overtones produced by a musical instrument is known as its:**
    (Average Rigor) (Domain III Skill 5)

    A. Timbre

    B. Chromaticity

    C. Resonant Frequency

    D. Flatness

Answer: A

To answer this question, you must know some basic physics vocabulary. "Timbre" is the combination of tones that make a sound unique, beyond its pitch and volume. (For instance, consider the same note played at the same volume, but by different instruments.) Answer (A) is therefore the only appropriate choice. "Resonant Frequency" is relevant to music, because the resonant frequency of a wave will give the dominant sound tone. "Chromaticity" is an analogous word to "timbre," but it describes color tones. "Flatness" is unrelated, and incorrect.

61. **The following statements about sound waves are true *except*:**
    (Average Rigor) (Domain III Skill 5)

    A. Sound travels faster in liquids than in gases.

    B. Sound waves travel through a vacuum.

    C. Sound travels faster through solids than liquids.

    D. Ultrasound can be reflected by the human body.

Answer: B

Sound waves require a medium to travel. The sound wave agitates the material, and this occurs fastest in solids, then liquids, then gases. Ultrasound waves are reflected by parts of the body, and this is useful in medical imaging. Therefore, the only correct answer is (B).

62. **The highest energy is associated with:**
    (Easy) (Domain III Skill 6)

    A. UV radiation

    B. Yellow light

    C. Infrared radiation

    D. Gamma radiation

Answer: D

To answer this question, recall the electromagnetic spectrum. The highest energy (and therefore frequency) rays are those with the lowest wavelength, i.e. gamma rays. (In order of frequency from lowest to highest are: radio, microwave, infrared, red through violet visible light, ultraviolet, X-rays, gamma rays.) Thus, the only possible answer is (D). Note that even if you did not remember the spectrum, you could deduce that gamma radiation is considered dangerous and thus might have the highest energy.

63. **Which of the following apparatus can be used to measure the wavelength of a sound produced by a tuning fork?**
    (Average Rigor) (Domain III Skill 7)

    A. A glass cylinder, some water, and iron filings

    B. A glass cylinder, a meter stick, and some water

    C. A metronome and some ice water

    D. A comb and some tissue

Answer: B

To answer this question, recall that a sound will be amplified if it is reflected back to cause positive interference. This is the principle behind musical instruments that use vibrating columns of air to amplify sound (e.g. a pipe organ). Therefore, presumably a person could put varying amounts of water in the cylinder, and hold the vibrating tuning fork above the cylinder in each case. If the tuning fork sound is amplified when put at the top of the column, then the length of the air space would be an integral multiple of the sound's wavelength. This experiment is consistent with answer (B). Although the experiment would be tedious, none of the other options for materials suggest a better alternative.

64. All of the following phenomena are considered "refractive effects" except for:
    (Rigorous) (Domain III Skill 8)

    A. The red shift

    B. Total internal reflection

    C. Lens dependent image formation

    D. Snell's Law

Answer: A

Refractive effects are phenomena that are related to or caused by refraction. The red shift refers to the Doppler Effect as applied to light when galaxies travel away from observers. Total internal reflection is when light is totally reflected in a substance, with no refracted ray into the substance beyond (e.g. in fiber optic cables). It occurs because of the relative indices of refraction in the materials. Lens dependent image formation refers to making images depending on the properties (including index of refraction) of the lens. Snell's Law provides a mathematical relationship for angles of incidence and refraction. Therefore, the only possible answer is (A).

65. A monochromatic ray of light passes from air to a thick slab of glass (n = 1.41) at an angle of 45° from the normal. At what angle does it leave the air/glass interface?
(Rigorous) (Domain III Skill 8)

    A. 45°

    B. 30°

    C. 15°

    D. 55°

Answer: B

To solve this problem use Snell's Law:
$n_1 \sin\theta_1 = n_2 \sin\theta_2$ (where n_1 and n_2 are the indexes of refraction and θ_1 and θ_2 are the angles of incidence and refraction).

Then, since the index of refraction for air is 1.0, we deduce:
1 sin 45° = 1.41 sin x
$x = \sin^{-1}((1/1.41) \sin 45°)$
x = 30°

This is consistent only with answer (B). Also, note that you could eliminate answers (A) and (D) in any case, because the refracted light will have to bend at a smaller angle when entering glass.

66. **Which of the following is *not* a legitimate explanation for refraction of light rays at boundaries between different media?**
    (Rigorous) (Domain III Skill 8)

    A. Light seeks the path of least time between two different points

    B. Due to phase matching and other boundary conditions, plane waves travel in different directions on either side of the boundary, depending on the material parameters

    C. The electric and magnetic fields become decoupled at the boundary

    D. Light rays obey Snell's law

Answer: C

Even if the exact implications of each explanation are not known or understood, answer C can be chosen due to its plain incorrectness. The other responses involve more or less fundamental explanations for the refraction of light rays (which are equivalent to plane waves) at media boundaries.

67. If an object is 20 cm from a convex lens whose focal length is 10 cm, the image is:
(Rigorous) (Domain III Skill 9)

    A. Virtual and upright

    B. Real and inverted

    C. Larger than the object

    D. Smaller than the object

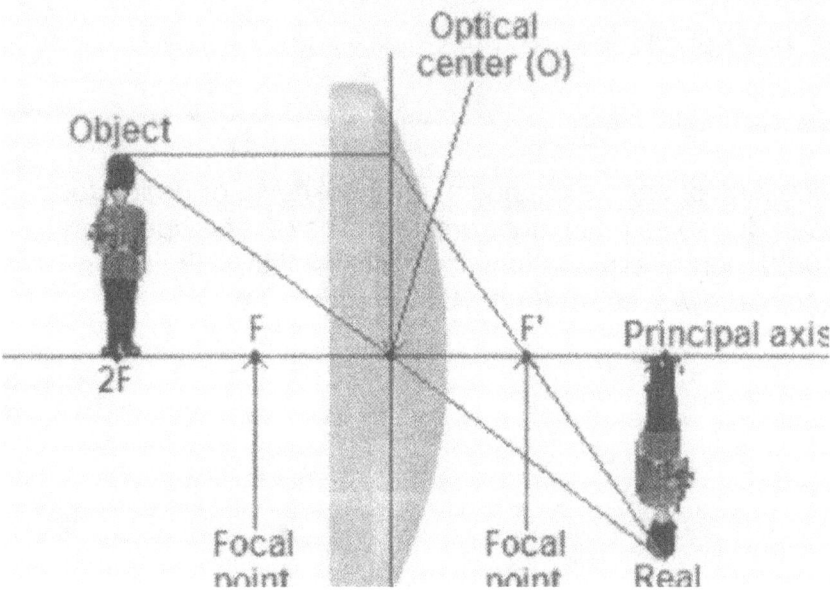

Answer: B

To solve this problem, draw a lens diagram with the lens, focal length, and image size.

The ray from the top of the object straight to the lens is focused through the far focus point; the ray from the top of the object through the near focus goes straight through the lens; the ray from the top of the object through the center of the lens continues. These three meet to form the "top" of the image, which is therefore real and inverted. This is consistent only with answer (B).

68. **Automobile mirrors that have a sign, "objects are closer than they appear" say so because:**
    (Rigorous) (Domain III Skill 10)

    A. The real image of an obstacle, through a converging lens, appears farther away than the object.

    B. The real or virtual image of an obstacle, through a converging mirror, appears farther away than the object.

    C. The real image of an obstacle, through a diverging lens, appears farther away than the object.

    D. The virtual image of an obstacle, through a diverging mirror, appears farther away than the object.

Answer: D

To answer this question, first eliminate answer choices (A) and (C), because we have a mirror, not a lens. Then draw ray diagrams for diverging (convex) and converging (concave) mirrors, and note that because the focal point of a diverging mirror is behind the surface, the image is smaller than the object. This creates the illusion that the object is farther away, and therefore (D) is the correct answer.

69. **An object two meters tall is speeding toward a plane mirror at 10 m/s. What happens to the image as it nears the surface of the mirror?**
    (Rigorous) (Domain III Skill 10)

    A. It becomes inverted.

    B. The Doppler Effect must be considered.

    C. It remains two meters tall.

    D. It changes from a real image to a virtual image.

Answer: C

Note that the mirror is a plane mirror, so the image is always a virtual image of the same size as the object. If the mirror were concave, then the image would be inverted until the object came within the focal distance of the mirror. The Doppler Effect is not relevant here. Thus, the only possible answer is (C).

70. **The boiling point of water on the Kelvin scale is closest to:**
    (Easy) (Domain IV Skill 1)

    A. 112 K

    B. 212 K

    C. 373 K

    D. 473 K

Answer: C

To answer this question, recall that Kelvin temperatures are equal to Celsius temperatures plus 273.15. Since water boils at 100°C under standard conditions, it will boil at 373.15 K. This is consistent only with answer (C).

71. **A calorie is the amount of heat energy that will:**
    (Easy) (Domain IV Skill 1)

    A. Raise the temperature of one gram of water from 14.5° C to 15.5° C.

    B. Lower the temperature of one gram of water from 16.5° C to 15.5° C

    C. Raise the temperature of one gram of water from 32° F to 33° F

    D. Cause water to boil at two atmospheres of pressure.

Answer: A

The definition of a calorie is, "the amount of energy to raise one gram of water by one degree Celsius," and so answer (A) is correct. Do not get confused by the fact that 14.5° C seems like a random number. Also, note that answer (C) tries to confuse you with degrees Fahrenheit, which are irrelevant to this problem.

72. **A temperature change of 40 degrees Celsius is equal to a change in Fahrenheit degrees of:**
    (Average Rigor) (Domain IV Skill 1)

    A. 40

    B. 20

    C. 72

    D. 112

Answer: C

To answer this question, recall the equation for Celsius and Fahrenheit:
°F = 9/5 °C + 32

Therefore, whatever temperature difference occurs in °C, it is multiplied by a factor of 9/5 to get the new °F measurement:
new°F = 9/5(old°C + 40) + 32
(whereas old°F = 9/5(old°C) + 32)
Therefore the difference between the old and new temperatures in Fahrenheit is 9/5 of 40, or 72 degrees. This is consistent only with answer (C).

73. **Solids expand when heated because:**
    (Rigorous) (Domain IV Skill 2)

    A. Molecular motion causes expansion

    B. PV = nRT

    C. Magnetic forces stretch the chemical bonds

    D. All material is effectively fluid

Answer: A

When any material is heated, the heat energy becomes energy of motion for the material's molecules. This increased motion causes the material to expand (or sometimes to change phase). Therefore, the answer is (A). Answer (B) is the ideal gas law, which gives a relationship between temperature, pressure, and volume for gases. Answer (C) is a red herring (misleading answer that is untrue). Answer (D) may or may not be true, but it is not the best answer to this question.

74. The number of calories required to raise the temperature of 40 grams of water at 30°C to steam at 100°C is:
(Rigorous) (Domain IV Skill 4)

    A. 7500

    B. 23,000

    C. 24,400

    D. 30,500

Answer: C

To answer this question, apply the equations for heat transfer due to temperature and phase changes:
$Q = mC\Delta T + mL$
where Q is heat; m is mass; C is specific heat; ΔT is temperature change; L is heat of phase change.

In this problem, we are trying to find Q, and we are given:
m = 40 g
C = 1 cal/g°C for water (this should be memorized)
ΔT = 70 °C
L = 540 cal/g for liquid to gas change in water (this should be memorized)

thus Q = (40 g)(1 cal/g°C)(70 °C) + (40 g)(540 cal/g)
Q = 24,400 cal
This is consistent only with answer (C).

75. Use the information on heats below to solve this problem. An ice block at 0° Celsius is dropped into 100 g of liquid water at 18° Celsius. When thermal equilibrium is achieved, only liquid water at 0° Celsius is left. What was the mass, in grams, of the original block of ice?

Given: Heat of fusion of ice = 80 cal/g
Heat of vaporization of ice = 540 cal/g
Specific Heat of ice = 0.50 cal/g°C
Specific Heat of water = 1 cal/g°C

(Rigorous) (Domain IV Skill 4)

A. 2.0

B. 5.0

C. 10.0

D. 22.5

Answer: D

To solve this problem, apply Conservation of Energy to the ice-water system. Any gain of heat to the melting ice must be balanced by loss of heat in the liquid water. Use the two equations relating temperature, mass, and energy:
$Q = m C \Delta T$ (for heat loss/gain from change in temperature)
$Q = m L$ (for heat loss/gain from phase change)
where Q is heat change; m is mass; C is specific heat; ΔT is change in temperature; L is heat of phase change (in this case, melting, also known as "fusion").

Then
$Q_{\text{ice to water}} = Q_{\text{water to ice}}$
(Note that the ice only melts; it stays at 0° Celsius—otherwise, we would have to include a term for warming the ice as well. Also the information on the heat of vaporization for water is irrelevant to this problem.)
$m L = m C \Delta T$
$x (80 \text{ cal/g}) = 100g \cdot 1\text{cal/g°C} \cdot 18°C$
$x (80 \text{ cal/g}) = 1800 \text{ cal}$
$x = 22.5 \text{ g}$
Only answer (D) matches this result.

TEACHER CERTIFICATION STUDY GUIDE

76. **Heat transfer by electromagnetic waves is termed:**
    (Easy) (Domain IV Skill 5)

    A. Conduction

    B. Convection

    C. Radiation

    D. Phase Change

Answer: C

To answer this question, recall the different ways that heat is transferred. Conduction is the transfer of heat through direct physical contact and molecules moving and hitting each other. Convection is the transfer of heat via density differences and flow of fluids. Radiation is the transfer of heat via electromagnetic waves (and can occur in a vacuum). Phase Change causes transfer of heat (though not of temperature) in order for the molecules to take their new phase. This is consistent, therefore, only with answer (C).

77. **A long copper bar has a temperature of 60°C at one end and 0°C at the other. The bar reaches thermal equilibrium (barring outside influences) by the process of heat:**
    (Average Rigor) (Domain IV Skill 5)

    A. Fusion

    B. Convection

    C. Conduction

    D. Microwaving

Answer: C

To answer this question, recall the different methods of heat transfer. (Note that since the bar is warm at one end and cold at the other, heat must transfer through the bar from warm to cold, until temperature is equalized.) "Convection" is the heat transfer via fluid currents. "Conduction" is the heat transfer via connected solid material. "Fusion" and "Microwaving" are not methods of heat transfer. Therefore the only appropriate answer is (C).

78. The First Law of Thermodynamics takes the form dU = dW when the conditions are:
    (Rigorous) (Domain IV Skill 7)

    A. Isobaric

    B. Isochloremic

    C. Isothermal

    D. Adiabatic

Answer: D

To answer this question, recall the First Law of Thermodynamics:
Change in Internal Energy = Work Done + Heat Added
dU = dW + dQ

Thus in the form we are given, dQ has been set to zero, i.e. there is no heat added. "Adiabatic" refers to a case where there is no heat exchange with surroundings, so answer (D) is the appropriate choice. "Isobaric" means at a constant pressure, "Isothermal" means at a constant temperature, and "Isochloremic" is an imaginary word, as far as I can tell.

It might be tempting to choose "Isothermal," thinking that no heat added would require the same temperature. However, work and internal energy changes can change temperature within the system analyzed, even when no heat is exchanged with the surroundings.

79. **What is temperature?**
    (Average Rigor) (Domain IV Skill 10)

    A. Temperature is a measure of the conductivity of the atoms or molecules in a material

    B. Temperature is a measure of the kinetic energy of the atoms or molecules in a material

    C. Temperature is a measure of the relativistic mass of the atoms or molecules in a material

    D. Temperature is a measure of the angular momentum of electrons in a material

Answer: B

Temperature is, in fact, a measure of the kinetic energy of the constituent components of a material. Thus, as a material is heated, the atoms or molecules that compose it acquire greater energy of motion. This increased motion results in the breaking of chemical bonds and in an increase in disorder, thus leading to melting or vaporizing of the material at sufficiently high temperatures.

80. **Bohr's theory of the atom was the first to quantize:**
    (Average Rigor) (Domain V Skill 1)

    A. Work

    B. Angular Momentum

    C. Torque

    D. Duality

Answer: B

Bohr was the first to quantize the angular momentum of electrons, as he combined Rutherford's planet-style model with his knowledge of emerging quantum theory. Recall that he derived a "quantum condition" for the single electron, requiring electrons to exist at specific energy levels.

81. Two neutral isotopes of a chemical element have the same numbers of:

(Easy) (Domain V Skill 2)

- A. Electrons and Neutrons
- B. Electrons and Protons
- C. Protons and Neutrons
- D. Electrons, Neutrons, and Protons

Answer: B

To answer this question, recall that isotopes vary in their number of neutrons. (This fact alone eliminates answers (A), (C), and (D).) If you did not recall that fact, note that we are given that the two samples are of the same element, constraining the number of protons to be the same in each case. Then, use the fact that the samples are neutral, so the number of electrons must exactly balance the number of protons in each case. The only correct answer is thus (B).

TEACHER CERTIFICATION STUDY GUIDE

82. **Which statement best describes why population inversion is necessary for a laser to operate?**
    (Rigorous) (Domain V Skill 2)

    A. Population inversion prevents too many electrons from being excited into higher energy levels, thus preventing damage to the gain medium.

    B. Population inversion maintains a sufficient number of electrons in a higher energy state so as to allow a significant amount of stimulated emission.

    C. Population inversion prevents the laser from producing coherent light.

    D. Population inversion is not necessary for the operation of most lasers.

Answer: B

Population inversion is a state in which there are a larger number of electrons in a particular higher-energy excited state than in a particular lower-energy state. When perturbed by a passing photon, these electrons may then emit a photon of the same energy (frequency) and phase. This is the process of stimulated emission, which, when population inversion is obtained, can produce something of a "chain reaction," thus giving lasers their characteristically monochromatic and highly coherent light.

83. **When a radioactive material emits an alpha particle only, its atomic number will:**
    (Average Rigor) (Domain V Skill 3)

    A. Decrease

    B. Increase

    C. Remain unchanged

    D. Change randomly

Answer: A

To answer this question, recall that in alpha decay, a nucleus emits the equivalent of a Helium atom. This includes two protons, so the original material changes its atomic number by a decrease of two.

TEACHER CERTIFICATION STUDY GUIDE

84. Ten grams of a sample of a radioactive material (half-life = 12 days) were stored for 48 days and re-weighed. The new mass of material was:
    (Rigorous) (Domain V Skill 3)

    A. 1.25 g

    B. 2.5 g

    C. 0.83 g

    D. 0.625 g

Answer: D

To answer this question, note that 48 days is four half-lives for the material. Thus, the sample will degrade by half four times. At first, there are ten grams, then (after the first half-life) 5 g, then 2.5 g, then 1.25 g, and after the fourth half-life, there remains 0.625 g. You could also do the problem mathematically, by multiplying ten times $(½)^4$, i.e. ½ for each half-life elapsed.

85. Which of the following pairs of elements are not found to fuse in the centers of stars?
    (Average Rigor) (Domain V Skill 7)

    A. Oxygen and Helium

    B. Carbon and Hydrogen

    C. Beryllium and Helium

    D. Cobalt and Hydrogen

Answer: D

To answer this question, recall that fusion is possible only when the final product has more binding energy than the reactants. Because binding energy peaks near a mass number of around 56, corresponding to Iron, any heavier elements would be unlikely to fuse in a typical star. (In very massive stars, there may be enough energy to fuse heavier elements.) Of all the listed elements, only Cobalt is heavier than iron, so answer (D) is correct.

86. **The constant of proportionality between the energy and the frequency of electromagnetic radiation is known as the:**
(Easy) (Domain V Skill 8)

      E. Rydberg constant

      F. Energy constant

      G. Planck constant

      H. Einstein constant

Answer: C

Planck estimated his constant to determine the ratio between energy and frequency of radiation. The Rydberg constant is used to find the wavelengths of the visible lines on the hydrogen spectrum.
The other options are not relevant options, and may not actually have physical meaning. Therefore, the only possible answer is (C).

87. A crew is on-board a spaceship, traveling at 60% of the speed of light with respect to the earth. The crew measures the length of their ship to be 240 meters. When a ground-based crew measures the apparent length of the ship, it equals:
    (Rigorous) (Domain V Skill 10)

    A. 400 m

    B. 300 m

    C. 240 m

    D. 192 m

Answer: D

To answer this question, recall that a moving object's size seems contracted to the stationary observer, according to the equation:
Length Observed = (Actual Length) $(1 - v^2/c^2)^{1/2}$
where v is speed of motion, and c is speed of light.

Therefore, in this case,
Length Observed = $(240 \text{ m}) (1 - 0.6^2)^{1/2}$
Length Observed = $(240 \text{ m}) (0.8) = 192 \text{ m}$

This is consistent only with answer (D). If you were unsure of the equation, you could still reason that because of length contraction (the flip side of time dilation), you must choose an answer with a smaller length, and only (D) fits that description. Note that only the dimension in the direction of travel is contracted. (The length in this case.)

88. If an object of length 5 meters (along the ŷ axis) is traveling at 99% of the speed of light in the x̂ direction, what is its length, as measured by an observer at the origin?
    (Rigorous) (Domain V Skill 10)

    A. 0 m

    B. 0.71 m

    C. 3.4 m

    D. 5 m

Answer: D

This problem, since it involves a speed close to c, must be treated relativistically. The answer may be quickly determined, however, by realizing that there is no observed length contraction perpendicular to the direction of travel. Thus, if an object is moving in the x direction, only its x dimension is contracted. As a result, there is no need to calculate the Lorentz factor; the answer is simply the proper length of the object (5 meters).

89. Which statement best describes a valid approach to testing a scientific hypothesis?
    (Easy) (Domain VI Skill 1)

    A. Use computer simulations to verify the hypothesis

    B. Perform a mathematical analysis of the hypothesis

    C. Design experiments to test the hypothesis

    D. All of the above

Answer: D

Each of the answers A, B and C can have a crucial part in testing a scientific hypothesis. Although experiments may hold more weight than mathematical or computer-based analysis, these latter two methods of analysis can be critical, especially when experimental design is highly time consuming or financially costly.

90. Which of the following is not a key purpose for the use of open communication about and peer-review of the results of scientific investigations?
    (Average Rigor) (Domain VI Skill 2)

    A. Testing, by other scientists, of the results of an investigation for the purpose of refuting any evidence contrary to an established theory

    B. Testing, by other scientists, of the results of an investigation for the purpose of finding or eliminating any errors in reasoning or measurement

    C. Maintaining an open, public process to better promote honesty and integrity in science

    D. Provide a forum to help promote progress through mutual sharing and review of the results of investigations

Answer: A

Answers B, C and D all are important rationales for the use of open communication and peer-review in science. Answer A, however, would suggest that the purpose of these processes is to simply maintain the status quo; the history of science, however, suggests that this cannot and should not be the case.

91. **Which description best describes the role of a scientific model of a physical phenomenon?**
    (Average Rigor) (Domain VI Skill 3)

    A. An explanation that provides a reasonably accurate approximation of the phenomenon

    B. A theoretical explanation that describes exactly what is taking place

    C. A purely mathematical formulation of the phenomenon

    D. A predictive tool that has no interest in what is actually occurring

Answer: A

A scientific model seeks to provide the most fundamental and accurate description possible for physical phenomena, but, given the fact that natural science takes an *a posteriori* approach, models are always tentative and must be treated with some amount of skepticism. As a result, A is a better answer than B. Answers C and D overly emphasize one or another aspect of a model, rather than a synthesis of a number of aspects (such as a mathematical and predictive aspect).

92. **Which of the following aspects of the use of computers for collecting experimental data is not a concern for the scientist?**
    (Rigorous) (Domain VI Skill 4)

    A. The relative speeds of the processor, peripheral, memory storage unit and any other components included in data acquisition equipment

    B. The financial cost of the equipment, utilities and maintenance

    C. Numerical error due to a lack of infinite precision in digital equipment

    D. The order of complexity of data analysis algorithms

Answer: D

Although answer D might be a concern for later, when actual analysis of the data is undertaken, the collection of data typically does not suffer from this problem. The use of computers does, however, pose problems when, for example, a peripheral collects data at a rate faster than the computer can process it (A), or if the cost of running the equipment or of purchasing the equipment is prohibitive (B). Numerical error is always a concern with any digital data acquisition system, since the data that is collected is never exact.

93. **What best describes an appropriate view of scale in scientific investigation?**
    (Rigorous) (Domain VI Skill 5)

    A. Scale is irrelevant: the same fundamental principles apply at all scale levels

    B. Scale has no bearing on experimentation, although it may have a part in some theories

    C. Scale has only a practical effect: it may make certain experiments more or less difficult due to limitations on equipment, but it has no fundamental or theoretical impact

    D. Scale is an important consideration both experimentally and theoretically: certain phenomena may have negligible effect at one scale, but may have an overwhelming effect at another scale

Answer: D

Although A is, in a sense, correct, in that the same principles do apply at all scales, their relative effect at various scales can vary drastically. Gravitational and quantum effects each may dominate at different scale levels, for example. As a result, both B and C are incorrect, since scale is a factor both theoretically and experimentally.

TEACHER CERTIFICATION STUDY GUIDE

94. Which phenomenon was first explained using the concept of quantization of energy, thus providing one of the key foundational principles for the later development of quantum theory?
    (Rigorous) (Domain VI Skill 6)

    A. The photoelectric effect

    B. Time dilation

    C. Blackbody radiation

    D. Magnetism

Answer: C

Although the photoelectric effect applied principles of quantization in explaining the behavior of electrons emitted from a metallic surface when the surface is illuminated with electromagnetic radiation, the explanation of the phenomenon of blackbody radiation, provided by Max Planck, was the first major success of the concept of quantized energy. Magnetism may be explained quantum mechanically, but such an explanation was not forthcoming until well after Planck's quantization hypothesis. Time dilation is primarily explained through relativity theory.

95. Which of the following is *not* an SI unit?
    (Average Rigor) (Domain VI Skill 8)

    A. Joule

    B. Coulomb

    C. Newton

    D. Erg

Answer: D

The first three responses are the SI (*Le Système International d'Unités*) units for energy, charge and force, respectively. The fourth answer, the erg, is the CGS (centimeter-gram-second) unit of energy.

96. Which statement best describes how significant figures must be treated when analyzing a collection of measured data?
   (Rigorous) (Domain VI Skill 9)

   A. Report all data and any results calculated using that data with the number of significant figures that corresponds to the least accurate number (i.e., the one with the fewest significant figures)

   B. Report all data and any results calculated using that data with the number of significant figures that corresponds to the most accurate number (i.e., the one with the most significant figures)

   C. Follow appropriate rules for the propagation of error, maintaining an appropriate number of significant figures for each particular set of data or calculated result

   D. Report all data with the appropriate number of significant figures, but ignore error when calculating results

Answer: C

It is critical, when analyzing data, to maintain the appropriate number of significant figures (a quantification of error) for each individual set of data or calculated result. To do this, certain rules must be followed to appropriately characterize the propagation of error from one or more data values, through various calculations, to the final result. The number of significant figures in the data may be different from the number in a calculated value, such as in an average or variance.

97. **Which of the following best describes the relationship of precision and accuracy in scientific measurements?**
    (Easy) (Domain VI Skill 11)

    A. Accuracy is how well a particular measurement agrees with the value of the actual parameter being measured; precision is how well a particular measurement agrees with the average of other measurements taken for the same value

    B. Precision is how well a particular measurement agrees with the value of the actual parameter being measured; accuracy is how well a particular measurement agrees with the average of other measurements taken for the same value

    C. Accuracy is the same as precision

    D. Accuracy is a measure of numerical error; precision is a measure of human error

Answer: A

The accuracy of a measurement is how close the measurement is to the "true" value of the parameter being measured. Precision is how closely a group of measurements is to the mean value of all the measurements. By analogy, accuracy is how close a measurement is to the center of the bulls-eye, and precision is how tight a group is formed by multiple measurements, regardless of accuracy. Thus, measurements may be very precise and not very accurate, or they may be accurate but not overly precise, or they may be both or neither.

98. **Which of the following experiments presents the most likely cause for concern about laboratory safety?**
    (Average Rigor) (Domain VI Skill 13)

    A. Computer simulation of a nuclear reactor

    B. Vibration measurement with a laser

    C. Measurement of fluorescent light intensity with a battery-powered photodiode circuit

    D. Ambient indoor ionizing radiation measurement with a Geiger counter.

Answer: B

Assuming no profoundly foolish acts, the use of a computer for simulation (A), measurement with a battery-powered photodiode circuit (C) and ambient radiation measurement (D) pose no particular hazards. The use of a laser (B) must be approached with care, however, as unintentional reflections or a lack of sufficient protection can cause permanent eye damage.

99. **If a particular experimental observation contradicts a theory, what is the most appropriate approach that a physicist should take?**
    (Average Rigor) (Domain VI Skill 22)

    A. Immediately reject the theory and begin developing a new theory that better fits the observed results

    B. Report the experimental result in the literature without further ado

    C. Repeat the observations and check the experimental apparatus for any potential faulty components or human error, and then compare the results once more with the theory

    D. Immediately reject the observation as in error due to its conflict with theory

Answer: C

When experimental results contradict a reigning physical theory, as they do from time to time, it is almost never appropriate to immediately reject the theory (A) *or* the observational results (D). Also, since this is the case, reporting the result in the literature, without further analysis to provide an adequate explanation of the discrepancy, is unwise and unwarranted. Further testing is appropriate to determine whether the experiment is repeatable and whether any equipment or human errors have occurred. Only after further testing may the physicist begin to analyze the implications of the observational result.

100. **Which situation calls might best be described as involving an ethical dilemma for a scientist?**
    (Rigorous) (Domain VI Skill 22)

    A. Submission to a peer-review journal of a paper that refutes an established theory

    B. Synthesis of a new radioactive isotope of an element

    C. Use of a computer for modeling a newly-constructed nuclear reactor

    D. Use of a pen-and-paper approach to a difficult problem

Answer: B

Although answer A may be controversial, it does not involve an inherently ethical dilemma, since there is nothing unethical about presenting new information if it is true or valid. Answer C, likewise, has no necessary ethical dimension, as is the case with D. Synthesis of radioactive material, however, involves an ethical dimension with regard to the potential impact of the new isotope on the health of others and on the environment. The potential usefulness of such an isotope in weapons development is another ethical consideration.

XAMonline, INC. 21 Orient Ave. Melrose, MA 02176

Toll Free number 800-509-4128

TO ORDER Fax 781-662-9268 OR www.XAMonline.com

WEST SERIES

PO#   Store/School:

Address 1:

Address 2 (Ship to other):

City, State Zip

Credit card number_____-_____-_____-_____   expiration_____

EMAIL _____

PHONE   FAX

ISBN	TITLE	Qty	Retail	Total
978-1-58197-638-0	WEST-B Basic Skills			
978-1-58197-609-0	WEST-E Biology 0235			
978-1-58197-693-9	WEST-E Chemistry 0245			
978-1-58197-566-6	WEST-E Designated World Language: French Sample Test 0173			
978-1-58197-557-4	WEST-E Designated World Language: Spanish 0191			
978-1-58197-614-4	WEST-E Elementary Education 0014			
978-1-58197-636-6	WEST-E English Language Arts 0041			
978-1-58197-634-2	WEST-E General Science 0435			
978-1-58197-637-3	WEST-E Health & Fitness 0856			
978-1-58197-635-9	WEST-E Library Media 0310			
978-1-58197-674-8	WEST-E Mathematics 0061			
978-1-58197-556-7	WEST-E Middle Level Humanities 0049, 0089			
978-1-58197-043-2	WEST-E Physics 0265			
978-1-58197-563-5	WEST-E Reading/Literacy 0300			
978-1-58197-552-9	WEST-E Social Studies 0081			
978-1-58197-639-7	WEST-E Special Education 0353			
978-1-58197-633-5	WEST-E Visual Arts Sample Test 0133			
	SUBTOTAL		Ship	$8.25
	FOR PRODUCT PRICES VISIT WWW.XAMONLINE.COM		TOTAL	

www.ingramcontent.com/pod-product-compliance
Lightning Source LLC
Chambersburg PA
CBHW080536300426
44111CB00017B/2749